THE ESSENCE OF

ENGINEERING THERMODYNAMICS

THE ESSENCE OF ENGINEERING SERIES

Published titles

The Essence of Solid-State Electronics
The Essence of Electric Power Systems
The Essence of Measurement
The Essence of Engineering Thermodynamics

Forthcoming titles

The Essence of Analog Electronics
The Essence of Circuit Analysis
The Essence of Optoelectronics
The Essence of Microprocessor Engineering
The Essence of Communications
The Essence of Power Electronics

THE ESSENCE OF ENGINEERING THERMODYNAMICS

James A. McGovern

Prentice Hall

LONDON NEW YORK TORONTO SYDNEY TOKYO
SINGAPORE MADRID MEXICO CITY MUNICH

First published 1996 by
Prentice Hall Europe
Campus 400, Maylands Avenue
Hemel Hempstead
Hertfordshire HP2 7EZ
A division of
Simon & Schuster International Group

Typeset in 10/12pt Times
by MHL Typesetting Ltd, Coventry

Printed and bound in Great Britain by
T.J. Press (Padstow) Ltd

Library of Congress Cataloging-in-Publication Data

McGovern, James A.
The essence of engineering thermodynamics / James A. McGovern.
p. cm. — (Essence of engineering)
Includes bibliographical references and index.
ISBN 0-13-518192-5
1. Thermodynamics. I. Title. II. Series.
TJ265.M38 1996
621.402'1—dc20 95-40396
CIP

British Library Cataloguing in Publication Data

A catalogue record for this book is available from the British Library

ISBN 0-13-518192-5

1 2 3 4 5 00 99 98 97 96

Contents

Preface

This book provides a first course in engineering thermodynamics that can be covered in 18 to 22 lectures. A background in mathematics, physics and chemistry, as would normally be possessed by engineering students taking such a first course, is assumed. It is expected, for instance, that the students are already familiar with Newton's laws and with chemical elements, such as carbon, and simple compounds, such as carbon dioxide. A basic knowledge of differential and integral calculus is assumed. The theory of thermodynamics is set out in a logical sequence and as concisely as possible. The objective has been to limit the material to that suitable for a first course, while dealing adequately with all the basics. Students are encouraged to consult other textbooks or references for a broader view of thermodynamics. However, they should find that this book provides them with a sufficient, readily understandable and appropriately rigorous treatment of the subject. Some lecturers will have supplemental material that they would like to include in the course. As the students will have adequate notes on the core material in this book, there should be time to provide such additional material. A set of tables of thermodynamic properties of water substance is included for the convenience of the students and the lecturer, as is a table giving ideal gas properties. A sufficient number of self-assessment questions are provided, where appropriate, at the end of the chapters and the answers are given at the back of the book. These questions are graded to help the students to develop progressively their knowledge and skills.

The book has evolved from the author's teaching over 14 years on an introductory thermodynamics course for engineering students at the University of Dublin, Trinity College. The author had found that most of the available textbooks that were suitable for recommending to the students were of a size that they found somewhat intimidating. They had to cope with a tightly packed curriculum and required concise material. The author provided this as comprehensive notes that were written out on an overhead projector. This generally worked quite well but the author was unhappy with the time taken up by the transcription. The notes were supplemented with handout sheets containing diagrams and with problems sheets that were given out at tutorials. As these built up it became a logical further step to prepare a complete volume of notes and this book.

Until very recently almost all engineering thermodynamics textbooks in the English language adopted the convention that energy transfer out of a system as work was positive. There has been some discussion at international conferences of the inherent inconsistency of this convention. The author considers that a compelling case has been made for taking work inputs as positive and thus using the same sign convention for heat and work. Some major textbooks (for instance, *Engineering Thermodynamics Work and Heat Transfer* by Rogers and Mayhew) have already broken with the old convention. The author also believes that the new convention is more easily adopted by students beginning thermodynamics than the old one.

The author wishes to acknowledge the help and encouragement given to him over many years by Professor W.G. Scaife, who also made numerous helpful comments about the draft manuscript. Thanks are also due to Professor Eugene Yantovski, Professor Pierre Le Goff and Dr Tony Tramschek for commenting on the draft manuscript.

Symbols

A	area
c	velocity, or specific heat capacity
const	constant
c.o.p.	coefficient of performance
CV	calorific value
D	diameter
E	energy, or total energy
e	specific energy
$\bar{e}$	molar energy
F	force
f	arbitrary function
f′	another arbitrary function
f″	another arbitrary function
g	acceleration due to gravity
H	enthalpy
h	specific enthalpy
i	general counter
L	length dimension
l	length
M	mass dimension
m	mass
$\dot{m}$	mass flow rate, or rate of change of mass
$\bar{m}$	molar mass
N	revolutions per unit time, or cycles per unit time
n	polytropic index, or number of moles, or number of items (systems, heat transfer interactions, work interactions, temperatures)
p	pressure
Q	electric charge dimension
Q	heat
$\dot{Q}$	heat transfer rate
q	heat transfer per unit mass
R	specific gas constant
$\bar{R}$	universal gas constant
S	entropy

s	specific entropy, or displacement
T	time dimension
T	absolute temperature
$\boldsymbol{T}$	torque
t	conventional temperature
$\boldsymbol{t}$	time
U	internal energy
u	specific internal energy
V	volume
v	specific volume
$\bar{v}$	molar volume
W	work, or non-heat energy transfer
$\dot{W}$	power
w	work per unit mass
X	arbitrary property
x	dryness fraction, or general x-ordinate
y	general y-ordinate
z	elevation, or general z-ordinate

Greek symbols

Δ	indicates a change of the following quantity
δ	indicates a minute amount or change of the following quantity
γ	adiabatic index (ratio of the specific heats)
η	efficiency
θ	angle
$\dot{\theta}$	rotational speed
ρ	density
ω	rotational speed

Subscripts

A+B	composite of systems A and B
atm	atmosphere
avg	average
b&s	boiler and superheater
bdry	boundary
Carnot	called after Carnot
c	refers to critical point
cmpr	compressor
cond	condenser
cr	creation
cyc	cycle
f	refers to saturated liquid
fg	refers to liquid–vapour saturation property difference
flow	associated with flow

fric	friction
fuel	fuel characteristic
g	refers to dry saturated vapour
H	refers to a heat source or sink at a high temperature
hp	heat pump
i	refers to saturated solid, or inside
in	into system
int	internal
irr	irreversible
k	kinetic
L	refers to a heat sink or source at a low temperature
liq	liquid
m	associated with mass
max	maximum
min	minimum
n	normal
net	refers to a net change, or a net effect, or a net value
o	outside
out	out of system
p	potential
p	at constant pressure
pmp	pump
Q	associated with heat transfer
r	radial, or reference
refr	refrigeration
rev	reversible
s	saturation property
sea	refers to the sea
surr	surroundings
sw	swept
sys	system
t	tangential
th	thermal
tp	triple point
trnsf	transfer
trnsp	transport
turb	turbine
use	useful
v	at constant volume or refers to volume
0	refers to the environment
$\rightarrow$	indicates the direction of a process or change
δA	corresponding to area δA

CHAPTER 1

Introduction

In this chapter the term *thermodynamics* is defined and its scope is explained. Some engineering application areas are briefly described. There is a quick review of the units that are used and of the need for dimensional consistency. Some explanation is given of how numerical data are represented in the book and in the tables in the appendices.

1.1 The scope of thermodynamics

Thermodynamics is the science of energy transformations involving work, heat and the properties of matter. This book is written for engineering students, but the theory of thermodynamics has a relevance that is much broader than the field of engineering with all its specializations. It is one of the widest of the sciences as it underlies many of the others.

Because thermodynamics is fundamentally concerned with work and energy, it includes the laws of mechanics. However, thermodynamics also deals with aspects of reality that are beyond the laws of mechanics.

Matter is composed of unimaginably vast numbers[1] of fundamental entities (molecules, atoms, nuclei, electrons and others) and each of these can have various types of energy. For example, at a given instant a molecule can have different components of kinetic energy in three orthogonal directions and different rotational kinetic energy components about three orthogonal axes. Much of the energy of a quantity of matter can be distributed in a random way over all the possible modes of energy possession of all its constituent entities. The situation is even more complicated than this as energy itself cannot be infinitely subdivided. The most minute entities can only undergo changes in their energy that represent a non-negligible part of the energy they possess. Quantum theory and statistics together with the laws of mechanics provide further insight into the types of phenomena that can occur when energy is minutely and randomly dispersed within matter. Whether or not quantum theory and statistical mechanics are explicitly used, thermodynamics deals

1. A rain drop with a volume of only 0.1 millilitres would contain about 3.3×10^{21} water molecules. The same tiny volume of air at standard atmospheric pressure (0.101 325 MPa) and 25 °C would contain about 2.5×10^{18} molecules (mostly nitrogen and oxygen).

with this dispersed energy that is possessed by matter and with exchanges of this energy between different bodies of matter. The science of thermodynamics can be applied to one minuscule entity, to any collection of entities, or even to the collection of entities that make up the universe.

Of all the fundamental laws of science there is only one that excludes the possibility of time proceeding backwards: the second law of thermodynamics. With a video camera it is possible to record the process of striking a match, lighting a candle, letting it burn for a time and blowing it out. The recording can be played back in reverse to show, supposedly, vapour igniting spontaneously about the wick, the candle unburning and the black char on the matchhead reacting rapidly with an inward flow of gas to form the original smooth coloured mass. In principle, the process could be instrumented to track the evolution of temperatures, forces, velocities and other measurable parameters. By playing back the instrument recordings in reverse it would be possible to demonstrate, supposedly, gases getting warmer as they approached the candle wick owing to heat transfer from the cooler surroundings, candle wax solidifying owing to heat transfer to the hotter flame and the matchstick pushing somebody's hand backwards owing to heat transfer from the surface of the matchbox. Although it is possible to describe and even show what would happen in the reverse process, it never happens like that in reality. The second law of thermodynamics, which is described in this book, is the formal statement of the law that excludes the possibility of processes such as the one described from happening in reverse. In fact it excludes the possibility of all real processes simultaneously occurring in reverse and thus excludes the possibility of time being reversed. This is a practical textbook on classical thermodynamics (quantum mechanics and statistical mechanics are not included) and it is not intended that it should be abstractly philosophical in its content. The student should be able to follow all steps of the reasoning, without exception. The concept of reversibility, which is introduced in Chapter 13, is one of the most important concepts in the science of thermodynamics. At the time of studying that chapter the student may find it useful to re-read the description of the candle process in this paragraph as an aid to understanding why certain processes cannot be reversed in the thermodynamic sense.

The laws of thermodynamics underlie our current scientific understanding of the universe. Astronomers have been able to apply these laws in tracing the development of the known universe from a very minute time interval after the Big Bang. When meteorologists forecast the weather they apply their knowledge of the laws of thermodynamics, which govern processes such as the generation of the winds and the formation of snow. Life itself is subject to the laws of thermodynamics and biologists take this into account in their work. Trees are subject to the laws of thermodynamics in using the Sun's energy to lift moisture from the ground to the tips of the tallest branches. Animal life converts some of the chemical energy of food to mechanical work and transfers heat to the surroundings. It is difficult to find an aspect of reality for which the science of thermodynamics does not have some relevance. Take the challenge: choose an arbitrary topic and see if you can link it to thermodynamics in some way. If you don't find a link straight away it may be worth while to put the idea in the back of your mind and to think about it afresh when you have the time.

There are two categories of perpetual motion machines that have taxed the imaginations of inventors for many centuries. An example of the first type would be a clockwork device that would periodically wind itself up using some of its own work output. It could thus operate indefinitely. An example of the second type would be a ship that would propel itself using only heat transfer from the sea. It would convert this heat transfer to work without requiring any heat sink[2] at a lower temperature. Both types of perpetual motion machine are excluded by the laws of thermodynamics. Whether such machines were to operate on principles that were based on mechanics, magnetics, lasers, electronics, microwaves, plasmas, chemical reactions, or strings and pulleys, the laws of thermodynamics would nonetheless apply. Engineers of all types need a basic understanding of these laws in order to avoid wasting time on defective concepts that might seem plausible to a person who had not been introduced to the theory of thermodynamics.

The concepts of temperature and heat are part of our everyday language. However, the science of thermodynamics is necessary in order that these concepts can be defined rigorously so that they can be used in a quantitative way.

1.2 Some engineering application areas

In the following sections some of the more common engineering applications of thermodynamics are described. The descriptions that are provided may be helpful for students who wish to understand how and where thermodynamics is applied in engineering practice. Traditionally, engineering thermodynamics has been closely associated with mechanical engineering. However, it is important to appreciate that the scope of engineering thermodynamics is much wider than the scope of a mechanical engineer. Refrigeration is usually produced mechanically by vapour compression and expansion, but it can also be produced by a machine that is mainly based on electronic principles, or on chemical principles. Electronic engineers or chemical engineers could be responsible for the design and operation of specialized refrigeration plants. Thermodynamics is important in the setting of concrete and in the behaviour of other materials used in civil engineering works. Building services engineers are responsible for the provision of heating, air-conditioning and ventilation, all of which involve the principles of thermodynamics.

The theory of thermodynamics that is set out in this textbook is intended to meet the needs of engineering students of all specializations at an introductory level. This theory can be expanded, by reference to more in-depth or more specialized works, to meet the needs of particular disciplines.

2. Some of the terms used in this paragraph, such as *heat* and *heat sink*, are explained later in the book. For the moment they can be interpreted as they are used in everyday language.

1.2.1 Power plants for electricity generation

Most of the electricity produced in the world is provided by generating stations in which water is evaporated at high pressure in a boiler to form steam. The steam drives turbines that in turn drive generators that provide the electricity. The low-pressure steam that leaves the turbines is condensed, rejecting heat to the environment, and the water is pumped back into the boiler. The working fluid, water, thus undergoes a cycle. The energy to evaporate the water can come from various sources, the most common being solid, liquid, or gaseous fossil fuels. Nuclear fission is another important source of energy for power generation. An indirect source that is commonly used is the exhaust gas of a gas turbine engine. A plant in which a gas turbine engine is used in this way in conjunction with a steam power plant is known as a combined cycle power plant. The most efficient power plants in the world, which are invariably combined cycle plants, convert a little more than 50% of the energy released from the fuel into electricity. The remaining fraction is rejected as heat to the environment. In less efficient power plants the energy rejected to the environment can be two-thirds or more of the fuel energy.

Engineers of various disciplines are involved in the design, construction and operation of steam power plants. These include mechanical engineers, electrical engineers, electronic engineers, civil engineers, computer engineers, control engineers and, for shipboard steam power plants, naval engineers. Chemical engineers are also interested in steam plants for the provision of process heating and power.

In Chapter 16 of this book the basic theory of the steam power plant cycle is presented. This makes use of the theory introduced in earlier chapters. It also provides a practical application of the thermodynamic tables for steam, as included in Appendix A.

1.2.2 Internal combustion engines

Spark ignition and compression ignition reciprocating engines for automotive use represent by far the greatest number of internal combustion engines produced. The gas turbine engines used in aircraft form another important category of internal combustion engines. Thermodynamics provides the theory on which the design of all these engines is based and provides the means by which improvements in performance can be sought.

The theory, at an introductory level, for spark ignition engines is presented in Chapter 17. This is the air standard Otto cycle.

1.2.3 Refrigeration and air-conditioning

In the developed world society is highly dependent on the technology of refrigeration and air-conditioning. Without refrigeration the nutritional value of food would be less and there would be much greater wastage of food in storage and in transit. Air-conditioning allows people to be comfortable and productive where it would otherwise be disagreeably hot or humid.

Most refrigeration and air-conditioning plants are based on a vapour compression cycle where the refrigerant is a two-phase working fluid. Liquid refrigerant is evaporated at low temperature and pressure while taking in energy as heat transfer from the cooled region. The vapour that leaves the evaporator is compressed to a high pressure and temperature in a compressor. It is then condensed, while rejecting heat to the environment. The liquid that leaves the condenser is expanded through a flow restriction and thus forms a mixture of liquid and vapour that enters the evaporator so that the cycle can continue. The cycle is similar to that of the steam power plant, but is in the reverse direction. The plant components and the overall cycle can be analyzed using the techniques described in this book for open systems and for heat engines.

1.2.4 Chemical process plants

Chemical thermodynamics is not covered in this textbook, but deserves a mention in this introduction. Chemical process plants include those for refining crude oil, producing pharmaceuticals and manufacturing diverse chemical substances such as plastics, adhesives, fertilizers and food additives. In the widest sense, plants for food processing and for brewing or distilling are also included. Desalination and water purification are other areas that can be regarded as chemical processing in the broad sense. Chemical reactions and mixing or separation processes involve energy transfer to or from the substances that are being processed. The laws of thermodynamics apply and the theory of thermodynamics provides the means by which chemical process plants can be designed to make the best use of energy.

1.3 Units and dimensions

The SI (Système International) system of units is used in this textbook and it is assumed that the student has already been introduced to this. Table 1.1 lists the units.

Unless otherwise stated, all pressures referred to in this book will be absolute pressures. *Absolute pressure* is the true force per unit area. In practice, pressure is commonly measured by pressure gauges that read gauge pressure. *Gauge pressure* is the amount by which a given pressure exceeds the pressure of the atmosphere. Atmospheric pressure varies depending on the weather and the altitude of the location where it is measured. A *standard atmosphere* is a pressure of 0.101 325 MPa. In many thermodynamics calculations the absolute pressure, rather than the gauge pressure, must be used.

The kilogram mole is the unit for quantity of substance regarded as a collection of entities at the atomic or molecular level. The entities would typically be atoms, molecules or ions (charged atoms or molecules). The particular type of entity has to be specified or implied from the context whenever the kilogram mole is used. The kilogram mole is defined in terms of the number of atoms of the reference substance carbon that would make up 1 kilogram. In nature some carbon atoms have

Table 1.1

Unit	*Abbreviation*	*Description*
Kilogram	kg	Unit of mass
Meter	m	Unit of length
Second	s	Unit of time
Newton	N	Unit of force
Pascal	Pa	Unit of pressure, 1 Pa = 1 N/m^2
Kelvin	K	Unit of temperature
Joule	J	Unit of energy, 1 J = 1 N m
Kilogram mole	kmol	Unit of quantity of specified entities, 1 kmol = (the same number of specified entities as there are atoms in 12 kg of carbon 12)
Watt	W	Unit of power, 1 W = 1 J/s
Radian	rad	Dimensionless unit of angular displacement
Non-SI supplementary units		
Bar	bar	1 bar = 10^5 Pa
Centimeter	cm	1 cm = 10^{-2} m
Degree Celsius	°C	Unit of temperature on the conventional temperature scale based on the freezing and boiling points of water {(1 °C temperature difference) = (1 K temperature difference)}
Gram	g	1 g = 10^{-3} kg
Litre	L	1 L = 10^{-3} m^3
Minute	min	1 min = 60 s
Hour	h	1 h = 3600 s
Mole	mol	1 mol = 10^{-3} kmol
Unit prefixes		
Kilo	k	10^3
Mega	M	10^6
Milli	m	10^{-3}

a significantly higher mass than the most common type by virtue of a higher number of neutrons in their nuclei. Therefore, it is necessary to specify that the carbon atoms in the definition of the kilogram mole are the most common type, known as carbon 12.

A *kilogram mole* is a quantity of specified entities equal to the number of atoms of carbon 12 that would make up 1 kilogram.

The number of entities in 1 kilogram mole is known as Avogadro's number and has the value 6.022×10^{26}. A *mole*[3] is defined as 10^{-3} kilogram moles.

All equations must be dimensionally correct. One way to check for this is to express all quantities in the equation in terms of the fundamental dimensions of mass **[M]**, length **[L]**, time **[T]** and electric charge **[Q]**, and to check that these fundamental dimensions balance in the equation. Mass, for instance, cannot be equated to time

3. The mole was originally defined in the context of a system of units in which the gram was the fundamental unit of mass.

or length. A more convenient method is to ensure that the units of all quantities balance over the entire equation: the units on the left-hand side must agree with those on the right-hand side. In solving problems, therefore, the units should be shown and used as a check on the dimensional validity of equations. This approach is adopted in all the examples in this book. A particularly important relation is

$$1\ \text{N} = 1\ \text{kg m/s}^2 \text{ (from Newton's second law).} \tag{1.1}$$

It is often necessary to use Equation (1.1) in checking equations for dimensional consistency.

EXAMPLE 1.1

Check the following equation for dimensional consistency by checking that the units balance:

$$1000\ [\text{kg m}^{-3}] \times 10\ [\text{m s}^{-2}] \times 100\ [\text{m}] \times 1\ [\text{m}^3] = 4\ [\text{kg}] \times 250\ [\text{J kg}^{-1}\ \text{K}^{-1}] \times 1000\ [\text{K}].$$

SOLUTION

Combining the numbers and units on each side of the equation

$$10^6\ \text{kg m}^2/\text{s}^2 = 10^6\ \text{J}.$$

From Newton's second law

$$1\ \text{N} = 1\ \text{kg m/s}^2.$$

Hence

$$1\ \text{kg m}^2/\text{s}^2 = 1\ \text{N m} = 1\ \text{J}.$$

Therefore, the equation can be written as

$$10^6\ \text{J} = 10^6\ \text{J}.$$

Answer

This is numerically and dimensionally correct.

The same combination of units as in this example could arise, for instance, in a situation where a given volume of water at a specified height in the Earth's gravitational field is allowed to pass to a lower level. The falling water could drive a turbine that would produce electricity to increase the temperature of a quantity of substance by means of a resistance heater.

1.4 Symbols

The list of the symbols used in this book to represent quantities such as pressure (p) and temperature (T) was given earlier. A dot over a symbol indicates either a time rate of change or a rate of occurrence. For instance, $\dot{m}$ can represent either a rate of change of the mass of a system or a mass flow rate (a rate of occurrence of mass transfer across a boundary).

1.5 Representation of numerical data

A symbol such as p represents a physical quantity, which is regarded as the product of a pure number and a unit that incorporates its dimensions. For example, the following equation defines a pressure.

$$p = 5 \text{ MPa}$$

where p represents the physical quantity pressure, 5 is a pure number and MPa is a unit that incorporates its physical dimensions ($[\mathbf{M}][\mathbf{L}^{-1}][\mathbf{T}^{-2}]$).

A physical quantity divided by its units is a pure dimensionless number. This provides a convenient way of representing physical data in tables or on graphs. In this book all tabulated or graphed values are pure dimensionless numbers. Tables that show the values of physical quantities have the form shown in Table 1.2.

Table 1.2

$\frac{p}{[\text{MPa}]}$
5
10
15
.
.
.

For instance,

$$\text{if } \frac{p}{[\text{MPa}]} = 5 \quad \text{then } p = 5 \text{ MPa.}$$

In a similar way, data can be graphed in dimensionless form by labelling the axes with the appropriate symbols for physical quantities divided by the units. See Figure 3.10 of Chapter 3.

1.6 Practical tips

- It is often necessary to use the absolute temperature rather than the conventional temperature in thermodynamics. The relationship between these scales is described fully in Chapter 4, section 4.4.
- It is often necessary to use the absolute pressure rather than the gauge pressure in thermodynamics.
- As a 'rule of thumb' an absolute temperature or pressure should be used if it is involved in multiplication, division, or exponentiation. Where a temperature difference is to be evaluated the same result will be obtained whether the two temperatures involved are in absolute or conventional units.

1.7 Summary

The scope of thermodynamics has been described and some examples of engineering application areas have been presented. Units and dimensions have been explained.
The student should be able to

- *define*
  — thermodynamics, absolute pressure, gauge pressure, a kilogram mole, a mole
- *describe*
  — the scope of thermodynamics in general terms
  — some engineering applications of thermodynamics
- *use*
  — the SI system of units for quantities encountered in thermodynamics
- *check*
  — equations for dimensional consistency by checking that the units balance
- *represent*
  — data in dimensionless form in tables and graphs.

1.8 Self-assessment questions

1.1 A room measures 3 m × 3.2 m × 2.9 m and contains air, which has a specific volume of 0.83 m^3/kg. Determine the mass of air present and its weight. Take the acceleration due to gravity as 9.81 m/s^2.

1.2 Check the following expression for dimensional consistency by ensuring that the units of the additive terms are consistent and evaluate it in kilowatts:

$$521 \left[\frac{\text{m}^2/\text{kg}}{\text{s}^3}\right] + 0.219 \left[\frac{\text{kJ}}{\text{kg}}\right] 1.255 \left[\frac{\text{kg}}{\text{s}}\right]$$

$$+ 135 \left[\frac{\text{Pa m}}{\text{s}}\right] 5.27 \left[\frac{\text{L}}{\text{kg}}\right] 7.72\ [\text{m}^2]\ 968 \left[\frac{\text{kg}}{\text{m}^2}\right]$$

CHAPTER 2

Systems, processes and interactions

In this chapter fundamental concepts relating to systems and processes are introduced. These include the concept of equilibrium and the idea of a cycle. Some of the definitions may seem very straightforward or even trivial. However, it will be found that a clear understanding of the fundamental definitions will simplify the more advanced concepts that are built upon them in later chapters.

Thermodynamics is a quantitative applied science. By making use of its principles, engineers can devise, analyze and evaluate useful plants and processes. As in other areas of science, there is a language to be learned so that concepts can be precisely described and so that practitioners can communicate effectively. The language of thermodynamics is partly verbal and partly diagrammatic. Some of the words used, like *system*, *heat* and *temperature*, are part of ordinary language but have a more restricted and precise meaning within the field of thermodynamics. In this chapter and in Chapters 3 and 4 many definitions are presented. The diagrammatic conventions of thermodynamics are introduced in the various figures.

2.1 Systems

A *system* is a region in space surrounded by a boundary that may be real or imaginary. Everything (and all space) that is not within a system's boundary is its *surroundings*.

A *subsystem* is any system that is contained within another system.

A *composite system* is any system that is composed of other systems. For instance, a boiled egg could be described as the composite of an egg–yoke system, an egg–white system and an egg–shell system. A composite system that consisted of six eggs could also be specified.

A *boundary* is a surface in space that has zero thickness and zero mass. Unless otherwise stated, it does not intervene in the forces between a system and its surroundings or in the transfer of mass or energy between them. However, in certain cases it may be convenient to associate specific constraints with a boundary. In reality these constraints must be realized by some material means. For example, the system of milk and air within a closed cardboard milk carton could be described as impermeably bounded. In reality the constraint is realized by a thin layer of polythene just outside the system boundary and on the inside of the carton. In diagrams, such

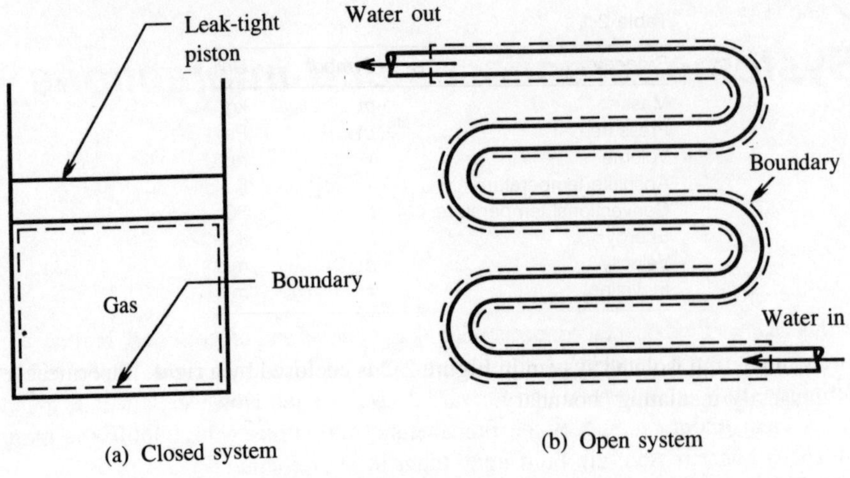

Figure 2.1 *Examples of a closed system and of an open system.*

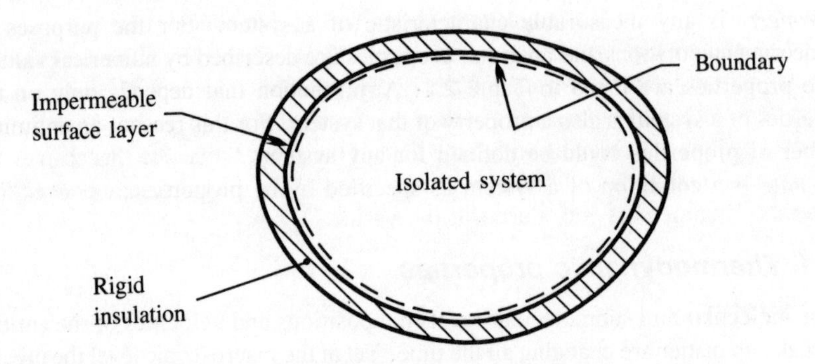

Figure 2.2 *An isolated system.*

as Figures 2.1 and 2.2, the boundary is represented by a dashed line and is shown slightly separated from any physical surface to which it corresponds.

An *impermeable boundary* is a boundary that does not permit matter to pass through.

A system is *closed* if matter cannot cross its boundary. Otherwise it is *open*. Figure 2.1 illustrates these two types of systems.

A *rigid boundary* is a boundary that resists any normal or shear forces exerted by the system without changing shape or size.

An *isolated boundary* is a boundary that permits neither matter nor energy[4] to pass through.

An *isolated system* is a system that is fully enclosed by an isolating boundary.

4. Energy is defined in Chapter 4.

Table 2.1

Property	*Symbol*	*Unit*
Mass	m	kg
Pressure	p	Pa
Volume	V	m^3
Absolute temperature	T	K
Conventional temperature	t	°C
Energy	E	J
Velocity	c	m/s
Elevation	z	m

For example, the isolated system in Figure 2.2 is enclosed by a rigid, impermeable and perfectly insulating[5] boundary.

2.2 Properties

A *property* is any measurable characteristic of a system. For the purposes of engineering calculations and analysis, properties are described by numerical values. Some properties are listed in Table 2.1. Any function that depends only on the properties of a system is also a property of that system. For this reason, an unlimited number of properties could be defined for any system.

A *state* is a condition of a system as specified by its properties.

2.2.1 Thermodynamic properties

At the molecular and submolecular levels the positions and velocities of the entities that make up matter are changing all the time. Yet at the macroscopic level the precise details of these ongoing changes are neither necessary nor useful[6] for the study of the characteristics of substances. A relatively small number of properties have been identified as useful for characterizing the state of substances in a quantitative way where the substances may undergo changes involving energy transformations, work or heat. Such properties are known as thermodynamic properties.

A *thermodynamic property* is a property that is independent of the individual positions or velocities of molecular or submolecular entities, is independent of the shape, position or velocity of a particular system or quantity of substance and does not require an external positional reference frame to define its value. Mass, pressure, volume and temperature are examples of thermodynamic properties. Some examples of properties that are not thermodynamic are the elevation of a system in the Earth's gravitational field; the velocity, kinetic energy or gravitational potential energy of

5. The term *adiabatic* would be more precise here than *perfectly insulating*. It is defined in Chapter 4, section 4.5.
6. Because of the impossibility of handling the vast quantities of data.

a system; and the diameter of a cylindrical system. These can be described as mechanical properties. A *mechanical property* is a property that describes the mechanical characteristics of a system.

A *thermodynamic state* is a condition of a system as specified by its thermodynamic properties. The complete specification of the state of a system may require the specification of non-thermodynamic properties in addition to thermodynamic properties.

A *force* is that which can cause a mass to accelerate.

A *macroscopic force* is a force defined in such a way that it is independent of the individual positions or motion of molecular or submolecular entities. A normal or shear stress multiplied by an area, a pressure multiplied by an area and the net force that acts on a system are examples.

2.2.2 Equilibrium

In mechanics a state of (Newtonian) equilibrium is defined within *Newton's first law*: a system remains in a state of rest or of uniform motion in a straight line unless acted upon by a force. In thermodynamics a more general definition, which also encompasses the Newtonian definition, is required. In thermodynamics strict Newtonian equilibrium is not demanded at the molecular or submolecular levels as this does not occur in reality. Instead it is recognized that a dynamic equilibrium state will eventually be established when any system is isolated from its surroundings. This state, known as an equilibrium thermodynamic state, will satisfy the special requirements of thermodynamic equilibrium for the substances the system contains and will meet the requirements of Newtonian equilibrium at the macroscopic level.

An *equilibrium thermodynamic state* (or *state of thermodynamic equilibrium*) of a system is a state such that if the system were isolated from its surroundings no change in its thermodynamic properties would occur. In particular, there are three less restrictive classes of equilibrium whose criteria must be satisfied by any system that is in thermodynamic equilibrium.

Mechanical equilibrium requires that no change would occur in the macroscopic forces acting on or within a system if it were isolated from its surroundings. This is Newtonian equilibrium for the system as a whole and for all subsystems in respect of the macroscopic forces that act on each of them and with respect to a reference frame attached to the system.

Thermal equilibrium requires that no temperature change would occur within any subsystem if the entire system were isolated from its surroundings.

Chemical equilibrium requires that no change would occur in the chemical composition of a system or of any of its subsystems if it were isolated from its surroundings.

Two systems are said to be in equilibrium with each other, or in mutual equilibrium, if the composite of the two systems would be in equilibrium if the two systems were brought together and enclosed by an isolating boundary. Systems can be in mutual thermodynamic equilibrium, or just in mutual mechanical equilibrium, thermal equilibrium, or chemical equilibrium.

For convenience in this book the word equilibrium on its own will mean thermodynamic equilibrium. Where necessary this will be made more explicit by a fuller description.

2.2.3 Extensive and intensive properties

An *extensive property* depends on the amount of substance present (e.g. m, V, E). An *intensive property* does not depend on the amount of substance present (e.g. p, T).

In most cases extensive properties have corresponding intensive forms, which are obtained by dividing the extensive property by the mass. The adjective *specific* is used to denote the resulting intensive properties, e.g. Table 2.2.

As an alternative to dividing by the mass, most extensive properties can also be divided by the number of kilogram moles present to yield intensive properties. The molecule would usually be the specified entity on which the kilogram mole was based. Table 2.3 includes three examples. Such properties are known as *molar* properties and are indicated by a bar over the symbol.

As a general rule upper case type is used for extensive properties and lower case type for intensive properties, but there are exceptions such as T, the absolute temperature, and R, the specific gas constant, which is described in Chapter 8.

A *point system* is a minute system, centred about a specified point, that is small enough that any differences between its intensive thermodynamic properties and those of an adjacent system of the same size can be neglected.

The *state at a point* is the state, as described by intensive properties, of a point system centred about that point.

Table 2.2

Property	*Symbol*	*Unit*
Specific volume	v	m^3/kg
Specific energy	e	J/kg

Table 2.3

Property	*Symbol*	*Unit*
Molar mass	$\bar{m}$	kg/kmol
Molar volume	$\bar{v}$	$m^3/kmol$
Molar energy	$\bar{e}$	J/kmol

2.2.4 Substance classification

In thermodynamics there are no restrictions on what a system may contain. For instance, it could contain a diesel electric generator or an entire industrial process plant. However, the focus is often on substances such as liquid water, steam, air, or the gases produced when a fuel is burned in air. The following terms are useful in describing substances.

A substance is *homogeneous* if it has a uniform physical structure and chemical composition.

A *phase* is a form in which a substance can exist that is homogeneous throughout. For example, water can exist in solid, liquid and vapour phases. As the word 'water' normally has the connotation of liquid, the term *water substance* will be used to refer to water without inferring that it is in a particular phase.

A *compressible substance* is a substance whose specific volume can vary. Gases are compressible. Liquids and solids are also compressible, although to a much lesser extent.

An *incompressible substance* is a substance whose specific volume is invariant. It is often possible to model liquids or solids as being incompressible for calculation purposes – as a result, the calculations are relatively straightforward. Some of engineering thermodynamics is concerned with systems that contain only an incompressible substance, but much of it is concerned with systems that contain a compressible substance in one or two phases.

A *fluid* is a substance that cannot remain in equilibrium while it is subjected to a shear force. Liquids and gases are fluids, as relative movement will occur within them whenever they are subjected to a shear force.

2.2.5 The state proposition

The following statement, known as the *state proposition*, is based on observation. It does not have the status of a law because it applies only to a specified type of substance. The number of independent intensive thermodynamic properties required to specify completely and uniquely the equilibrium thermodynamic state of a homogeneous compressible substance is two.

Hence, for a compressible substance in a given phase all equilibrium thermodynamic properties can be expressed in terms of any two such properties that are independent. An *equation of state* describes the interrelationships between certain properties over a range of equilibrium states, e.g. $f(p, v, T) = 0$. The ideal gas equation, Equation (2.1), is an instance. It is described in Chapter 8.

$$pv - RT = 0 \tag{2.1}$$

2.2.6 Point and path functions

There are two types of mathematical functions whose change depends on the values of specified independent variables. They provide a useful basis for distinguishing between quantities that are properties of a system, or of a substance, and quantities that are not. This was a major issue in the early days of thermodynamics, as heat was thought, incorrectly, to be a property.

A mathematical *point function* is a function of specified independent variables whose change depends only on the initial and final values of the variables.

For instance, the difference in elevation of the Earth's surface between two points is a function of the latitude and longitude only. If the coordinates of these points are specified then the difference in elevation has a unique value.

A mathematical *path function* is a function of specified independent variables whose change depends on the way in which the variables change in relation to one another between their initial and final values.

A knowledge of the initial and final values of the independent variables is not sufficient to specify the change in the function. A path function is therefore not a point function. An example would be the distance travelled between two points on the Earth's surface where no particular route is specified. This would depend on the path taken between the points.

As a property can have only one value at any given state of a system, all properties are point functions of a set of independent properties that describe the state, by definition. For systems in general and states that are not necessarily equilibrium states, the number of independent properties necessary to specify a given state may be indeterminately large. However, according to the state proposition, any intensive thermodynamic property of a system in equilibrium that contains only a homogeneous compressible substance is a point function of just two independent intensive thermodynamic properties.

2.2.7 Sources and sinks

A *source* is a system that can provide a specified extensive property. Mass and energy are examples of properties that could be provided by a source. Thus, a water source provides mass in the form of water. A *heat source* provides energy as heat transfer. A *mechanical work source* provides energy as mechanical work. The qualifying words 'heat' and 'mechanical work' describe how the energy is provided by these particular types of source and do not imply that heat and work are properties. Mechanical work is defined in Chapter 3. Energy and heat are defined in Chapter 4.

A *sink* is a system that can accept a specified extensive property. Two common examples would be a *heat sink* and a *mechanical work sink*. These accept the property energy. An electric power sink accepts energy as a rate of transfer of electrical energy at its boundary. A mass sink accepts the thermodynamic property mass.

2.3 Processes

A *process* is a change from one state to another that is undergone by a system or a quantity of matter, e.g. a heating process where the temperature of a system increases (for instance, water in a kettle), or compression of a gas where the volume of a sealed container is reduced (as in a gas spring).

A *flow process* occurs when the state of a quantity of matter changes as it passes through an open system. An example would be the heating of water in an open system such as that shown in Figure 2.1(b). A flow process can be regarded as the process undergone by a point system that passes through an open system. The point system has the same velocity as the matter at all the fixed points that constitute its path through the open system.

An *equilibrium process* is a process during which a system or quantity of matter remains in equilibrium while its state changes.

A *non-equilibrium process* is a process where a system does not remain in equilibrium as its state changes. A real process that occurs in a finite time will always be a non-equilibrium process, but a process where the deviations from equilibrium are negligible is described as a *quasi-equilibrium process*. Many real processes can be regarded as quasi-equilibrium processes for the purposes of thermodynamic analysis or design.

A *state diagram* (or *process diagram*) is a graph of one thermodynamic property against another on which equilibrium states can be represented as points and equilibrium processes can be represented as continuous curves (or paths); for example, Figure 2.3 shows a particular process on two different state diagrams.

A *cycle* is a process or series of processes that starts and ends at the same state of a system. Figure 2.4 illustrates a cycle. A cycle can be regarded as made up of a series of infinitesimal processes. For any property, X, the sum of all the infinitesimal changes that occur over any cycle is zero. This is expressed mathematically, using the symbol for the cyclic integral, in Equation (2.2):

$$\oint dX = 0 \quad \text{for any cycle (}X\text{ could be } p, T, V, E, \text{ etc.).} \tag{2.2}$$

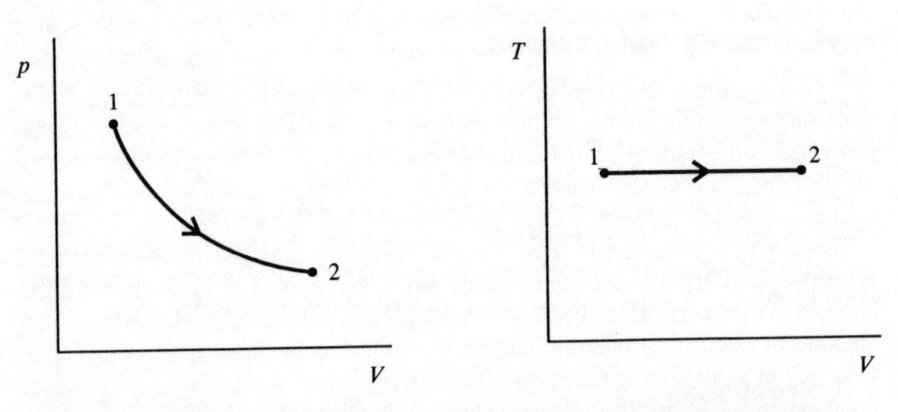

Figure 2.3 An equilibrium process shown on p–V and T–V diagrams. This could be a slow expansion of a gas in an uninsulated closed system. As the volume of the gas increases, the pressure decreases and the temperature remains unchanged.

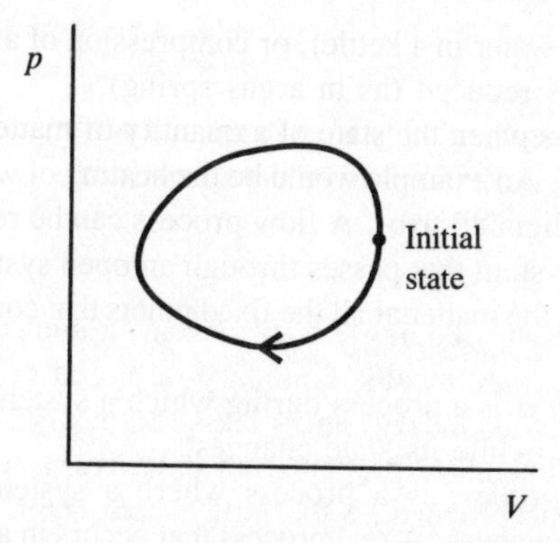

Figure 2.4 *An equilibrium cycle on a p–V diagram.*

2.4 Interactions

An *interaction* is said to occur or to exist between two systems when one system influences or sustains the state of the other system.

An interaction between systems may or may not involve a change in the state of one or both of the systems. For instance, there is a *force interaction* between a gas system and a rigid cylinder that contains it, even if both systems are in equilibrium. There will be a *thermal interaction* between a hot block of metal and a cold block of metal when they are brought into contact. In this case there will be a change in the state of both systems.

2.4.1 Transfer and transport

In thermodynamics, the term *transfer* indicates that an extensive property passes from one system to another. When matter enters or leaves a system, this can be described as the transfer of matter into or out of the system. Work and heat, which are described in Chapters 3 and 4, are modes of energy transfer between systems. In these cases the property energy passes from one system to another.

The term *transport* is used to describe any transfer between systems of an extensive property where the transfer can be quantified as the product of a mass transfer and an intensive property at the boundary that separates the systems. Matter transports amounts of extensive properties in much the same way as a train transports people. One way to visualize the transport of a property across a boundary is to consider point systems at a specified intensive state passing through the boundary of an open system. Energy and momentum are examples of properties that can be transported across a system boundary.

2.5 Summary

Some fundamental concepts relating to systems and processes have been introduced and many fundamental definitions have been provided.

The student should be able to

- *define*
  — a system, the surroundings, a subsystem, a composite system, a boundary, an impermeable boundary, a closed system, an open system, a rigid boundary, an isolating boundary, an isolated system, a property, an extensive property, an intensive property, a state, a thermodynamic property, a thermodynamic state, a mechanical property, a force, a macroscopic force, an equilibrium thermodynamic state, a state of thermodynamic equilibrium, mechanical equilibrium, thermal equilibrium, chemical equilibrium, when two systems are in mutual equilibrium, an extensive property, an intensive property, a point system, the state at a point, homogeneous, a phase, a compressible substance, an incompressible substance, a fluid, an equation of state, a point function, a path function, a source, a sink, a process, a flow process, an equilibrium process, a non-equilibrium process, a quasi-equilibrium process, a state diagram, a process diagram, a cycle, an interaction, transfer, transport

- *explain*
  — the term *molar*, a force interaction, a thermal interaction, a heat source, a mechanical work source, a heat sink, a mechanical work sink, the cyclic integral of a property

- *state*
  — Newton's first law, the state proposition

- *sketch*
  — systems and system boundaries
  — equilibrium processes and cycles on state diagrams.

CHAPTER 3

Work

The concept of work is introduced in this chapter. Mechanical work is defined. Two types of mechanical work at the boundary of a system are described. Electric work and electric power transfer at a boundary are discussed. The sign convention for work is introduced. Friction and fluid friction are described.

3.1 Vector and scalar quantities

A *vector* is a quantity that requires both a magnitude and a direction in space to describe it. A displacement (or movement) of the tip of a welding electrode attached to a robot arm is an example. It could be described by a magnitude in metres and a direction with respect to a reference frame, e.g. a displacement of 0.85 metres at an angle of 32° above the horizontal and 17° east of north. Velocity and acceleration are further examples of vector quantities.

A *scalar* is a quantity that requires only a magnitude to describe it. It is not associated with any particular direction. Mass, for instance, is a scalar quantity. It can be described by a magnitude in kilograms.

3.2 Work

3.2.1 Mechanical work

Force is a vector quantity. Its magnitude is related to the acceleration of a mass by Newton's second law. Its direction is the direction in which a point mass would accelerate if it were subjected to only that force. A force can be described by a magnitude in newtons and a direction with respect to a reference frame. The *line of action of a force* is an imaginary straight line that extends to infinity in both directions and is collinear with the force vector.

Mechanical work is the movement of the point of action of a force through a distance in the direction of the force. The work done on a system that is subjected to a force and undergoes a displacement is defined as the product of the force and the displacement along the line of action of the force.

If a constant force F acts on a system that is displaced through a distance s in

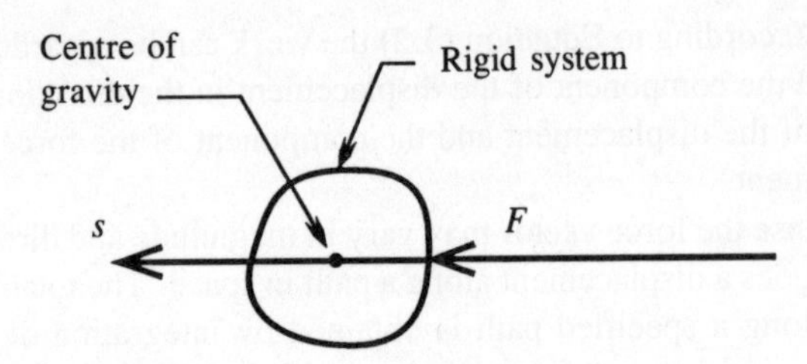

Figure 3.1 *Diagram of a force vector F and a displacement vector s in the direction of the force. The line of action of the force passes through the centre of gravity of the system and the displacement is that of the point where the force acts.*

the direction of the force, as in Figure 3.1, the work done on the system is given by Equation (3.1). The SI unit of work is the joule (1 J = 1 N m).

$$W = Fs. \tag{3.1}$$

It is not necessarily the case that the displacement at the point of action of a particular force is in the direction of the force. The system may already have been moving before the force was applied and there may be other forces acting to constrain or influence the motion at the point of action of the force of interest. A good example would be a roller coaster or big dipper. The force of gravity that acts on a roller coaster carriage always acts vertically downwards. Yet, due to the action of the force of gravity and the motion the carriage already has at any given point, the carriage descends and climbs, accelerates and decelerates, leans and turns, and even loops the loop, as it is constrained to do by the fixed tracks.

Figure 3.2 shows a situation where a small displacement δs (which, in the limit, can be infinitesimal) is not in the same direction as the force F, although it has a component in that direction. The incremental amount of work done on the system that is subjected to the force and undergoes this small displacement is given by Equation (3.2).

$$\delta W = F\delta s \cos\phi \tag{3.2}$$

where ϕ is the angle between the force vector and the displacement vector in the

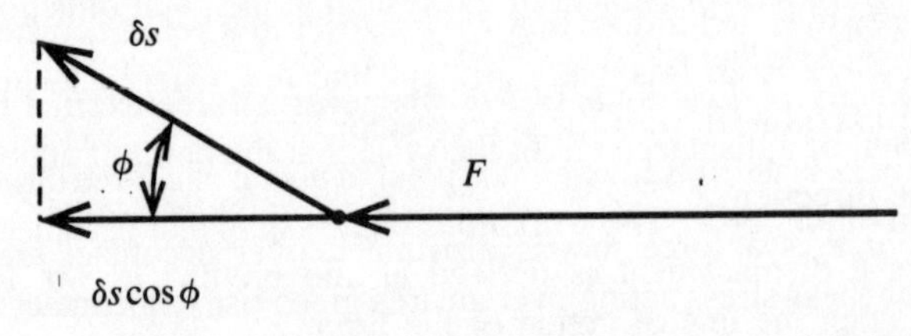

Figure 3.2 *Diagram of a force vector F, a displacement vector δs and the component of the displacement vector in the direction of the force δs cos ϕ.*

plane of both. According to Equation (3.2) the work can be regarded as the product of the force and the component of the displacement in the direction of the force or as the product of the displacement and the component of the force in the direction of the displacement.

In a general case the force vector may vary in magnitude and direction as its point of action undergoes a displacement along a path in space. The total work for a finite displacement along a specified path is obtained by integration of Equation (3.2):

$$W = \int F \cos\phi \; ds. \tag{3.3}$$

This can be evaluated if the path followed by the point of action of the force is known and if the force vector is known as a function of the position along the path. Work is transitory and mechanical work exists only while a force acts at a point where a displacement is occurring.

If the force F has a constant magnitude and the angle ϕ is zero for the entire displacement path, then Equation (3.3) reduces to the form of Equation (3.1). If the angle between the displacement vector and the force vector in Figure 3.2 is 90°, then, from Equation (3.2), the incremental work is zero as cos90° is zero. This means that there is no work if there is no component of displacement in the direction of the force (or no component of the force in the direction of the displacement).

If the angle (measured either anticlockwise or clockwise) between the displacement vector and the force vector is greater than 90° and less than or equal to 180°, then the component of the displacement vector along the line of action of the force has the opposite direction to the force vector. From Equation (3.2), the work done on the system that is subjected to the force is then negative. Newton's third law allows the meaning of this negative work to be understood. According to Newton's third law, whenever a force F is exerted on a system by a second system, an equal and opposite force $-F$ is exerted on the second system by the first system. If the work done on the first system is given by $F\delta s \cos\phi$, then the work done on the second system for the same displacement is given by $-F\delta s \cos\phi$. Therefore, negative work done on a system is equivalent to positive work done on the other system involved in the work interaction if the displacement is the same for both. Based on the preceding mathematical treatment, the following two statements can be made to summarize when mechanical work is done on a system and when it is done by a system.

Mechanical work is done on a system whenever a force exerted by the surroundings and a component of a displacement of the system at the point of action of the force have the same direction.

Mechanical work is done by a system whenever a force exerted by the system and a component of a displacement of the system at the point of action of the force have the same direction.

A *contact force* is a force that is distributed over a contact area. A tensile, compressive or shear stress acting over an area gives rise to a contact force. A *body force* is a force that is distributed over three-dimensional space within a system. Gravitational forces and electromagnetic forces are examples. A *point force* is a force that acts at a point. Generally this is a theoretical ideal, but is useful as a model in many practical situations. A point force can be regarded as the limiting case of either a contact force or a body force, where the area or the volume respectively is infinitesimal. For calculation purposes the force of gravity, which is a body force, can usually be regarded as a point force that acts at the centre of gravity of the system being analyzed. Analysis of other body forces is beyond the scope of this book.

Mechanical work is not the only type of work that can occur. Electric work and magnetic work are other types, but a detailed treatment of these is not part of this course. The only type of non-mechanical work likely to be encountered at the level of this introductory thermodynamics course is the electric work done over a period of time within an electric cable that passes through a system boundary. This type of work is discussed in section 3.3.3. A general definition of work is given in Chapter 14, section 14.8. It depends on the property entropy, which is introduced in that chapter, and on the property energy which is defined in Chapter 4, section 4.2. Ultimately, the general definition of work depends on the definition of mechanical work, as energy is defined in terms of mechanical work.

3.2.2 Power

Power is a rate of work. The SI unit of power is the watt (1 W = 1 J/s). Power is represented by the symbol $\dot{W}$:

$$\dot{W} = \frac{\mathrm{d}W}{\mathrm{d}t}. \tag{3.4}$$

If an amount of work W is done over a period of time Δt, then the average rate of work for that time interval, or the average power, is given by

$$\dot{W}_{\mathrm{avg}} = \frac{W}{\Delta t}. \tag{3.5}$$

3.3 Work at the boundary of a system

Contact forces can give rise to work at the boundary of a system. Figure 3.3 shows a force F that acts at the boundary of a system while its point of action on the boundary moves through a small distance δs, which, in the limit, can be regarded as infinitesimal.

The increment of work δW on the system for this process is given by Equation (3.2). Such work can occur where the boundary itself moves or where a point system that is subjected to the force moves through a boundary that is fixed.

3.3.1 Shear or shaft work

Shear work or *shaft work* is mechanical work that involves only a displacement in the plane of the boundary of a system. Shear work commonly occurs in association

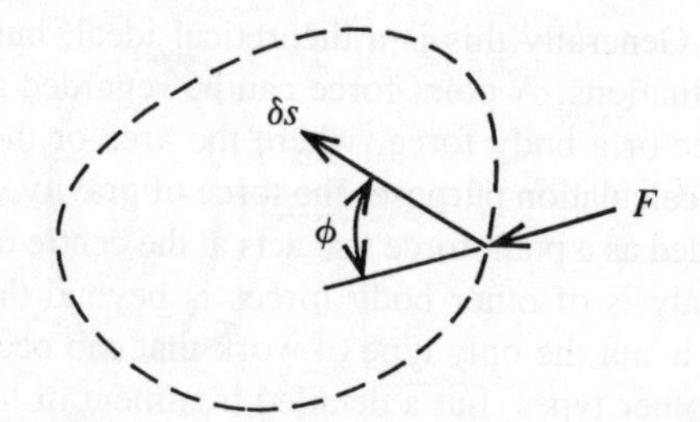

Figure 3.3 *Work at the boundary of a system.*

with the rotational displacement of a shaft, but can also occur in association with a linear displacement.

Figure 3.4 shows a hydraulic crane that lifts a weight. The system boundary has been drawn in such a way that it cuts through the horizontal lifting arm. A shear force, equal to the weight mg (neglecting the weight of the arm outside the boundary), acts downwards on the system in the plane of the boundary within the lifting arm. A shear force acts upwards on the surroundings. When the weight is being lifted,

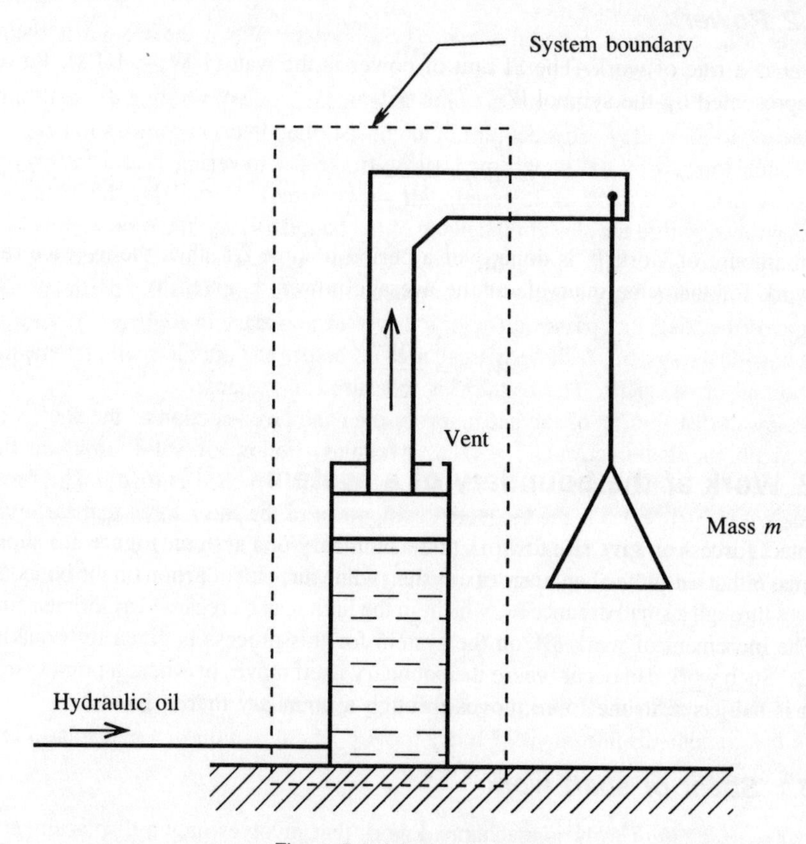

Figure 3.4 *A hydraulic crane.*

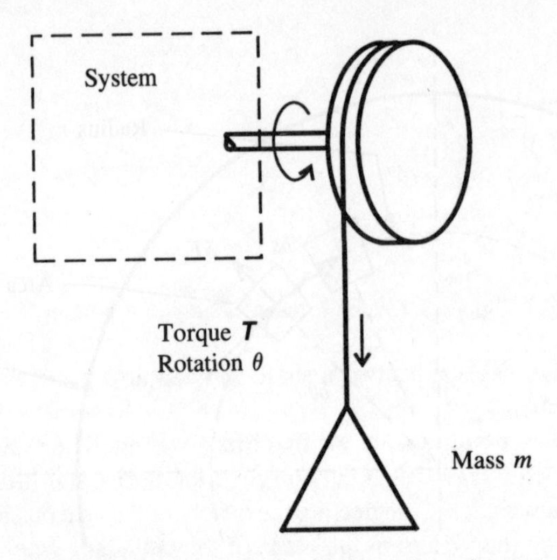

Figure 3.5 *A rotating shaft at the boundary of a system.*

work is done on the surroundings and by the system. When the weight is being lowered, work is done on the system.

Figure 3.5 shows a system that contains a device capable of rotating the shaft that is attached to the pulley. This device could consist of an internal combustion engine and clutch for raising the weight and a disc brake for lowering it at a controlled speed. When the weight is lowered, the shear forces exerted by the system's surroundings, within the shaft in the plane of the boundary, act in the same direction as the displacement. Therefore, the work is done on the system. The shear work can be examined in more detail with reference to Figure 3.6. This is a partial cross-section of the shaft that passes through the system boundary in Figure 3.5, viewed from outside the system. It shows a small area δA before and after a small incremental rotation $\delta\theta$ of the shaft. The rotation is measured in radians.

The exact distribution of the shear forces over the cross-section of the shaft will depend on its shape (circular, square, irregular, hollow or solid), and on the characteristics and distribution of the materials from which it is made. The force δF_t shown in Figure 3.6 is the tangential component of the shear force that acts over the small area δA. Any radial shear force such as δF_r is not involved in the work process as there is no displacement in the radial direction. From Equation (3.2)

$$\delta W_{\delta A} = \delta F_t \delta s \tag{3.6}$$

where $\delta W_{\delta A}$ is the shear work corresponding to area δA for rotation $\delta\theta$.

As the angular displacement $\delta\theta$ tends to zero, the direction of the displacement vector δs becomes the tangential direction and

$$\delta s = r\delta\theta \tag{3.7}$$

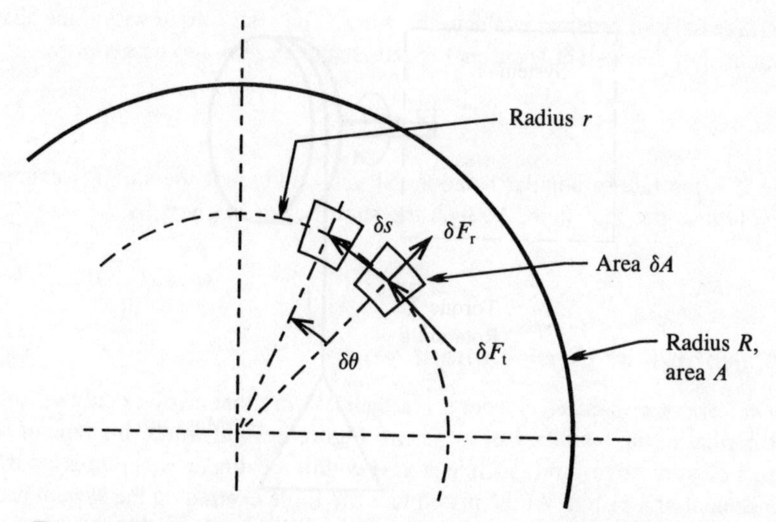

Figure 3.6 Diagram showing a small rotation $\delta\theta$ of the shaft in Figure 3.5.

where $\delta\theta$ is expressed in radians. Therefore,

$$\delta W_{\delta A} = r\delta F_t \delta\theta = \delta \boldsymbol{T} \delta\theta \tag{3.8}$$

where $\delta\boldsymbol{T}$ is the torque about the centre of rotation of the shaft due to the shear force δF_t over area δA.

The total torque is the sum of the amounts of torque for all the small areas making up the cross-sectional area A of the shaft:

$$\boldsymbol{T} = \Sigma\delta\boldsymbol{T}. \tag{3.9}$$

The work corresponding to the entire cross-sectional area for the rotation increment $\delta\theta$ is given by

$$\delta W = \Sigma\delta W_{\delta A} = \Sigma\delta\theta\delta\boldsymbol{T} = \delta\theta\Sigma\delta\boldsymbol{T} = \boldsymbol{T}\delta\theta. \tag{3.10}$$

For a finite rotation θ of the shaft, the work is obtained by integration of Equation (3.10):

$$W = \int_0^\theta \boldsymbol{T}\,\mathrm{d}\theta. \tag{3.11}$$

If the torque in the shaft is constant, then

$$W = \boldsymbol{T}\theta. \tag{3.12}$$

It is seen that a detailed knowledge of the shear force distribution within the shaft is not necessary in order to evaluate the work. The total torque within the shaft is sufficient. For a constant rotational speed ω, the *shaft power* is given by

$$\dot{W} = \boldsymbol{T}\dot{\theta} = \boldsymbol{T}\omega \tag{3.13}$$

where ω is the rate of angular rotation (SI units: rad/s). If the speed is expressed in revolutions per unit time, N, then the shaft power is given by

$$\dot{W} = 2\pi N\boldsymbol{T} \tag{3.14}$$

3.3.2 Normal or displacement work

Normal work or *displacement work* is mechanical work that involves only a displacement normal to the boundary of a system. Figure 3.7 illustrates this type of work where a compressible fluid (such as a gas) within a cylinder is compressed by the displacement of a piston. At the piston face the force exerted on the system has the same direction as the displacement. Therefore, work is done on the system.

From Equation (3.3) and as the force vector and the displacement vector are collinear

$$W = \int F \, \mathrm{d}s = \int pA \, \mathrm{d}s \tag{3.15}$$

where p is the absolute pressure of the system and, as $\delta V = -A\delta s$,

$$W = -\int p \, \mathrm{d}V. \tag{3.16}$$

Because a fluid in equilibrium exerts a force that is normal to any infinitesimal area of the boundary, the expression $W = -\int p \, \mathrm{d}V$ applies irrespective of the shape of the boundary. The integral can be evaluated only if p is known as a function of V for the entire process. The equilibrium normal work is thus a path function of the properties p and V.

Figure 3.8 shows how the absolute pressure varies with the volume of a closed system that contains only a compressible fluid for an arbitrary equilibrium process from state 1 to state 2. The hatched area underneath the curve is the integral of the

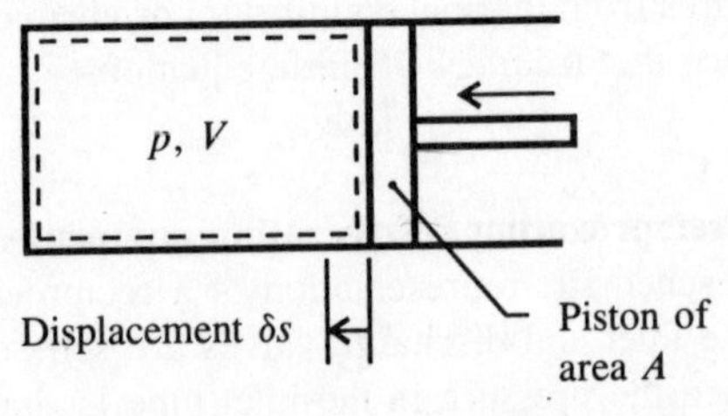

Figure 3.7 *Normal work (or displacement work) at the boundary of a system.*

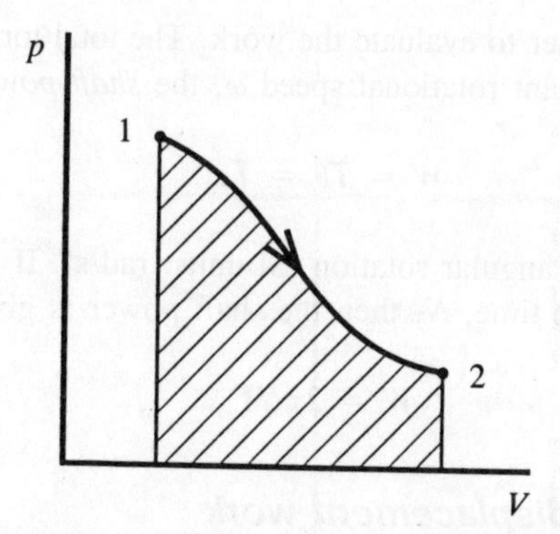

Figure 3.8 *An equilibrium displacement work process of a closed system.*

absolute pressure with respect to the volume, Equation (3.17):

$$(\text{area below } p{-}V \text{ process curve}) = \int_1^2 p \, dV. \qquad (3.17)$$

If the process proceeds in the direction of increasing volume, the hatched area is positive as the volume change is positive and the absolute pressure is positive. If the process proceeds in the direction of decreasing volume, the area underneath the curve will have a negative value.

Combining Equations (3.16) and (3.17), the relationship between the work and the area below the process curve is given by

$$W = -(\text{area below } p{-}V \text{ process curve}). \qquad (3.18)$$

In using this equation it must be kept in mind that the hatched area is positive where the process is in the direction of increasing volume and negative where the process is in the direction of decreasing volume. Equation (3.18) is useful in cases where the variation of pressure with volume can be measured for a closed system and where the system can be regarded as in mechanical equilibrium during the process. The work can then be determined by calculating the area under the process curve, as in Example 3.1. Any departure from mechanical equilibrium will reduce the accuracy of Equations (3.16) and (3.18) in predicting the normal or displacement work. It is notable that departures from thermal equilibrium or chemical equilibrium do not in themselves influence the accuracy of these equations.

EXAMPLE 3.1

Case study: a reciprocating refrigeration compressor

Figure 3.9 is a schematic representation of a reciprocating refrigeration compressor. The inlet and discharge valves are self-acting. The inlet valve opens when the pressure in the inlet pipe is slightly higher than that in the cylinder. The discharge valve opens when the pressure in the

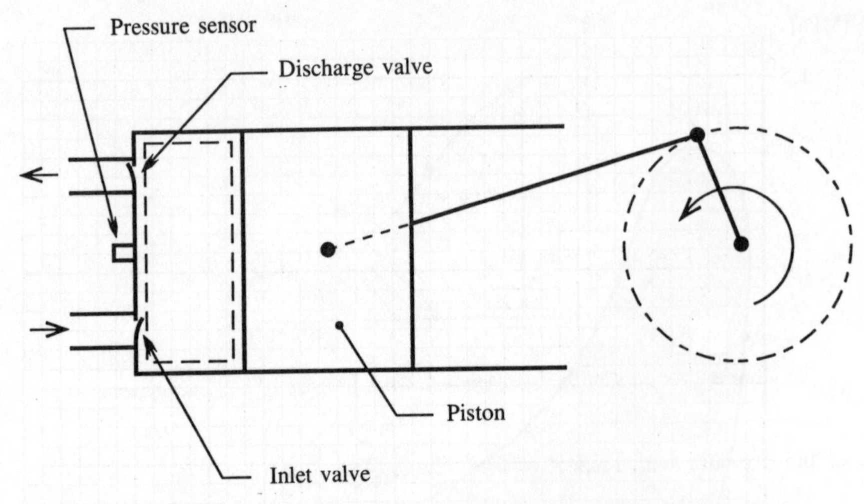

Figure 3.9 *Schematic representation of a reciprocating refrigeration compressor.*

cylinder is slightly higher than the pressure in the discharge pipe.

Figure 3.10 shows a graph of pressure versus volume for the refrigerant vapour within the cylinder. The pressure was measured at normal running speed over one rotation of the crankshaft by means of a pressure sensor, and the instantaneous values were captured and stored by a data acquisition system. The volume was measured indirectly by logging the point at which the piston reached the top-dead-centre position (the position where the volume within the cylinder is minimum) and measuring the speed of rotation of the shaft. From these measurements and the dimensions of the compressor, the instantaneous volume within the cylinder was calculated. Such a diagram is known as an indicator diagram. An arrow has been added to the diagram to indicate the anticlockwise sense of the process. Dynamic measurements were also made of the positions of the inlet and discharge valves over one cycle. From these measurements the points of opening and closing of each valve have been marked on the diagram. The inlet valve was open from point 1 to point 2, while the discharge valve was open from point 3 to point 4. Between points 2 and 3 and between points 4 and 1, the cylinder was closed. The compressor speed was 592 revolutions per minute.

Determine the swept volume of the compressor (the difference between the maximum cylinder volume and the minimum cylinder volume) from the horizontal scale. By counting the small boxes formed by the grid that are within the cycle, calculate the work per cycle. The area within the cycle is equivalent to the integral in Equation (3.17) and equals the negative of the work done on the system per cycle. Calculate also the average rate at which work is done on the system. This is known as the indicated power of the compressor.

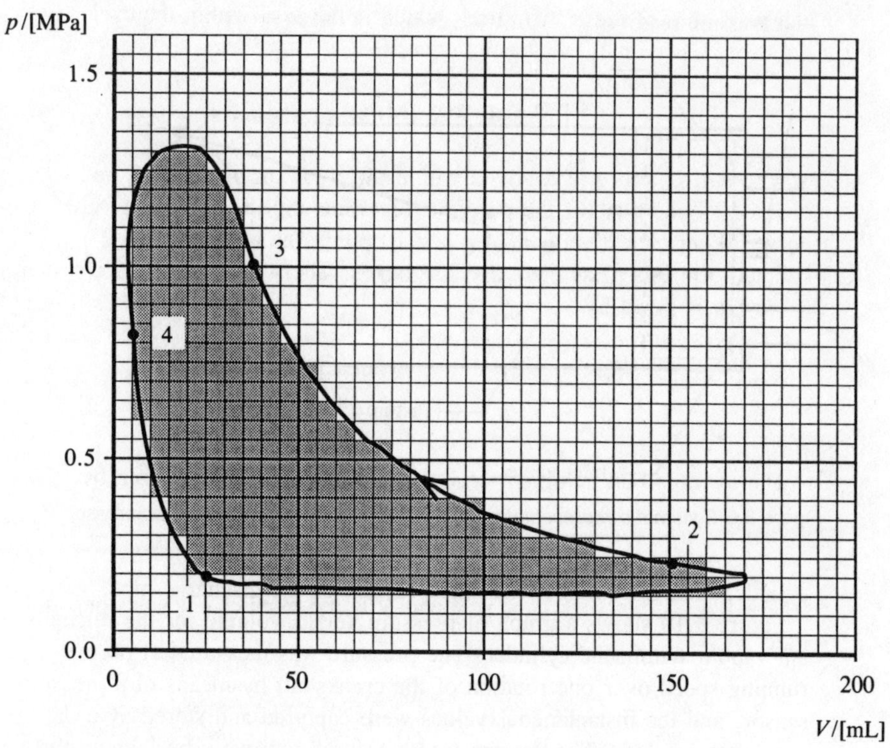

Figure 3.10 *An experimentally determined indicator diagram for a reciprocating refrigeration compressor.*

Examine the compression process while the cylinder is closed between points 2 and 3 to see if it has the form pV^n = const. Evaluate the exponent n.

SOLUTION

From the indicator diagram the maximum cylinder volume is V_{max} = 170 mL and the minimum cylinder volume is V_{min} = 4 mL. The swept volume is given by

$$V_{sw} = V_{max} - V_{min}$$
$$= (170 - 4)\text{ mL} = 166\text{ mL}.$$

The area within the cycle is calculated by counting the small boxes formed by the grid within the cycle. The area within the cycle is equivalent to the area under the upper process curve minus the area under the lower process curve. The area under the upper curve is negative, as the process is in the negative volume direction. The area under the lower curve is positive, as the process is in the positive volume direction. The algebraic sum of these two areas, which is the area within the cycle, is negative. Each small box has an area of

$$-(5 \text{ [mL]} \times 0.05 \text{ [MPa]}) = -0.25 \text{ J}.$$

Where half or more of a box is within the $p-V$ curve, this can be counted as one box for the purposes of estimating the area within the curve. Where less than half of a box is within the curve, it is not counted. On this basis there are 265 boxes. The work done on the system per cycle is given by

$$W_{cyc} = -\int p \, dV = -(\text{area within cycle})$$
$$= -265 \times (-0.25 \text{ [J]}) = 66.25 \text{ J}.$$

The average rate at which work is done by the piston is given by

$$\dot{W} = W_{cyc} \times N$$

where W_{cyc} is the work per cycle and N is the number of cycles per unit time.

$$\dot{W} = 66.25 \text{ [J]} \times \frac{592}{60} \text{ [s}^{-1}\text{]} = 653.7 \text{ W}.$$

The mean torque for the cycle is given by

Answer

$$\boldsymbol{T} = \frac{\dot{W}}{\omega} = \frac{653.7 \text{ [W]}}{(592/60) \times 2\pi \text{ [rad s}^{-1}\text{]}} = 10.54 \text{ N m}.$$

From the $p-V$ diagram at point 2

$$V_2 = 150.5 \text{ mL}$$
$$p_2 = 0.24 \text{ MPa}.$$

From the $p-V$ diagram at point 3

$$V_3 = 38 \text{ mL}$$
$$p_3 = 1.005 \text{ MPa}.$$

The exponent n is evaluated as follows:

$$p_2 V_2^n = p_3 V_3^n$$

$$\left(\frac{V_3}{V_2}\right)^n = \frac{p_2}{p_3}$$

$$n \log\left(\frac{V_3}{V_2}\right) = \log\left(\frac{p_2}{p_3}\right)$$

$$n = \frac{\log(p_2/p_3)}{\log(V_3/V_2)}$$

Answer

$$n = \frac{\log(0.24/1.005)}{\log(38/150.5)} = 1.04.$$

The accuracy of the relationship pV^n = const in predicting the pressure at some particular point in the process, given the volume, can now be checked. For instance, at a volume of 75 mL the pressure shown in Figure 3.10 is 0.501 MPa. The value given by the relationship with the exponent n is as follows:

$$p = p_2\left(\frac{V_2}{75\ [\text{mL}]}\right)^{1.04}$$

$$p = 0.24\ [\text{MPa}]\left(\frac{150.5\ [\text{mL}]}{75\ [\text{mL}]}\right)^{1.04} = 0.495\ \text{MPa}.$$

Answer
Therefore, the relationship predicts the pressure at a volume of 75 mL with good accuracy. The relationship is thus a good representation of the measured data.

3.3.3 Electric power

In many practical situations, energy is supplied to a system over a period of time as electric power rather than as mechanical power. The laws of thermodynamics do not exclude the possibility of fully converting electric power to mechanical power, or the reverse, by means of ideal motors and generators. Even in practice, very high conversion efficiencies (e.g. 98% or better) can be achieved. Therefore, electric power transfer across a boundary is taken to be equivalent to a rate of energy transfer as mechanical shaft work. Figure 3.11 illustrates electric power transfer at a system

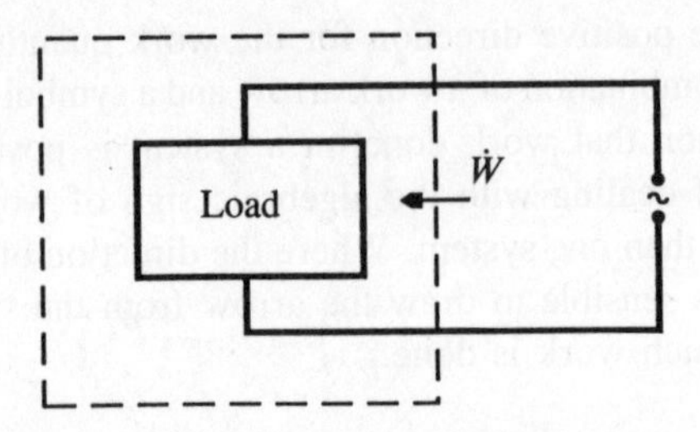

Figure 3.11 *Electric power transfer at the boundary of a system.*

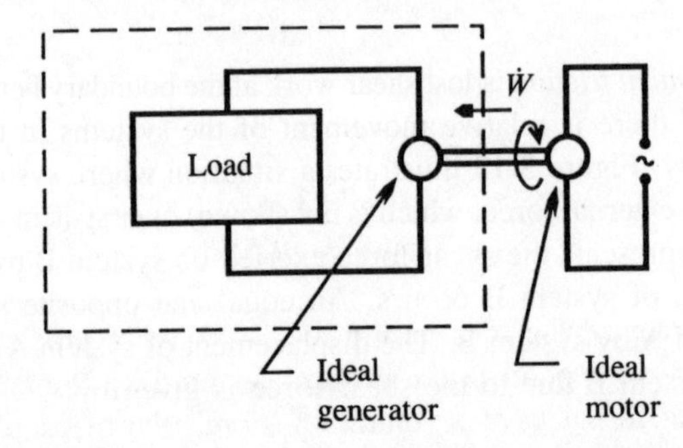

Figure 3.12 *Equivalent shaft power transfer at the boundary of a system corresponding to a transfer of electric power.*

boundary. Figure 3.12 illustrates the equivalent mechanical power transfer, which has the same value.

3.4 The sign convention for work

In situations where attention is focused on one particular system, it is convenient to define positive and negative work with respect to that system. *Work done on a system* is taken to be *positive* with respect to the system. *Work done by a system* is taken to be *negative*. This convention can be overridden by explicitly indicating the positive direction for work if this proves more convenient. Whenever a work quantity appears within a thermodynamic relation without qualification, the convention will be taken to apply.

> **Caution:** For historical reasons, many textbooks in the English language use exactly the opposite sign convention for work. In applying thermodynamic relations from diverse sources it is important to establish which sign convention has been adopted in each.

In diagrams in this textbook a special arrowhead (—•) is used to identify work or a work rate. This is a solid arrowhead with five corners. See, for instance, Figures 3.11 and 3.12. If the arrow is labelled with a symbol, then the direction of the arrow is to be taken as the positive direction for the work quantity represented by that symbol. Thus, the combination of a work arrow and a symbol takes precedence over the normal convention that work done on a system is positive. This is also the preferred method of dealing with the algebraic sign of work when the focus of attention is on more than one system. Where the direction of an energy transfer as work is known, it is sensible to draw the arrow from the system that does work to the system on which work is done.

3.5 Friction

Friction or *mechanical friction* is lost shear work at the boundary between two systems that arises where there is relative movement of the systems in the plane of their common boundary. Figure 3.13 illustrates a situation where system B moves (due to the action of an external force, which is not shown) over system A, which is fixed. The symbol F_B represents the shear force exerted on system B by system A while a displacement s_B of system B occurs. An equal and opposite shear force F_A is exerted on system A by system B. The displacement of system A, s_A, is zero. The work done on system B due to the shear force is given by

$$W_B = -F_B s_B. \tag{3.19}$$

This is negative. In other words, work is done by system B. The work done on system A is given by

$$W_A = F_A s_A = 0. \tag{3.20}$$

Thus, the interaction between the two systems involves work done by system B, but no work done on system A. The lost work is called *friction work*. The description of the interaction in terms of mechanics is complete, but a fuller description can be given in terms of thermodynamics. This is done in Chapter 10, section 10.3.

$$W_{fric} = -W_B - W_A = F_B s_B. \tag{3.21}$$

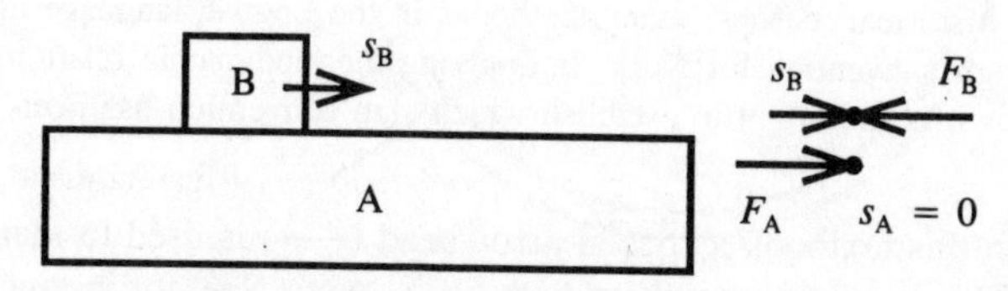

Figure 3.13 *Friction work at the common boundary of systems A and B.*

3.6 Fluid friction

Fluid friction is lost shear work that arises whenever a fluid system is subjected to shear forces.

Figure 3.14 illustrates a device that can be used to measure fluid friction. The fluid is placed in the space between a rotating drum and a fixed cylinder that surrounds it. The drum is driven by a motor at a constant speed and the torque applied to it can be measured. At the surface of the drum there is a shear force between the drum and the fluid. Work is done by the drum and on the fluid. At any other radius within the fluid there is also a circumferential shear force that is equal to the torque divided by the radius. However, the tangential fluid velocity varies from the constant value at the surface of the drum to zero at the inside surface of the enclosing cylinder. No work is done on the fixed cylinder. In terms of mechanics there is a rate of work input to the fluid system but there is no work output rate. There is thus a loss of shear work. This loss, which is due to fluid friction, is distributed within the volume of the fluid system. The lost work can be described as *friction work*. The shear forces within the fluid are known as *viscous forces*.

3.7 Summary

Mechanical work has been defined and explained. Mechanical work at the boundary of a system has been categorized as shear (or shaft) work or normal (or displacement) work. Expressions have been presented and explained for the evaluation of these types of work. Electric work and power transfer have been discussed. The sign convention for work has been presented. A description in terms of mechanics has been given of friction and fluid friction.

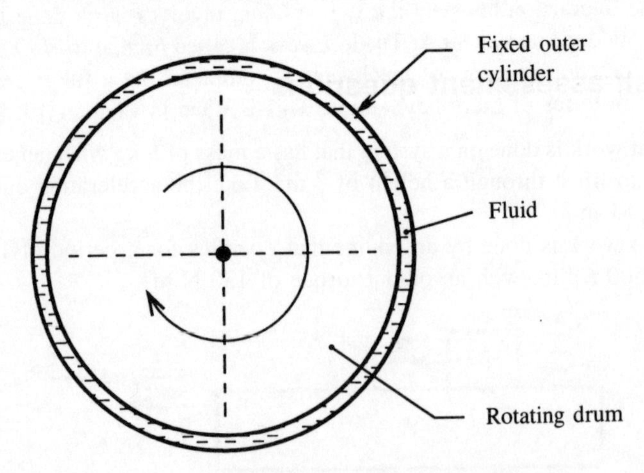

Figure 3.14 *View from above of a device, known as a viscometer, that can measure fluid friction.*

The student should be able to

- *define*
  - — a vector, a scalar, the line of action of a force, when mechanical work is done on a system, when mechanical work is done by a system, a contact force, a point force, a body force, power, shear work, shaft work, normal work, displacement work, positive work, negative work, friction, mechanical friction, fluid friction
- *describe by an example*
  - — friction, fluid friction
- *evaluate*
  - — mechanical work as the product of a force and a displacement
  - — mechanical work as the product of a torque and an angular displacement
- *represent*
  - — equilibrium normal or displacement work as the negative of an area under a process curve on a $p-V$ process diagram
- *distinguish between*
  - — mechanical work done on a system and mechanical work done by a system
- *state*
  - — the sign convention for work
- *explain*
  - — normal work as a function of pressure and volume, the equivalence of mechanical power and electric power.

3.8 Self-assessment questions

3.1 What work is done on a system that has a mass of 5 kg when an external force acts to lift it through a height of 3 m? Take the acceleration due to gravity as 9.81 m/s^2.

3.2 What work is done by an engine that operates for a period of 15 minutes at 3500 r.p.m. with an output torque of 130 N m?

CHAPTER 4

Mass, energy, temperature and heat

The principle of conservation of mass and the principle of conservation of energy are introduced in this chapter. The fundamental concepts of energy, temperature and heat are defined and explained. The sign conventions for heat transfer and for energy transfer in general across the boundary of a system are introduced.

4.1 The principle of conservation of mass

The *principle of conservation of mass* states that mass cannot be created or destroyed, but only changed in form. It is a special case of a more general principle: that of the conservation of mass–energy. Einstein's equation $E = mc^2$ relates mass and energy (c is the velocity of light in this equation). However, in almost every case in engineering thermodynamics no detectable conversion between mass and energy occurs. Two notable exceptions are the cases of nuclear fission and fusion reactions, where mass is converted to energy. These are very specific cases and very special means are necessary to achieve conversion between mass and energy. Normally, however, the principle of conservation of mass holds.

Mass, which is an extensive thermodynamic property, is conserved, but this is not the case for all extensive thermodynamic properties. The principle of conservation of mass is based on observation. It is a simple principle, but it is not trivial.

4.2 Energy

Energy is the capacity to do mechanical work, whether at the macroscopic, molecular or subatomic level. The SI unit of energy is the joule. The types of energy that are encountered most often in engineering situations are described in the following sections.

4.2.1 Kinetic energy

Kinetic energy is the energy possessed by a mass due to its velocity. Suppose a system has a mass m and a velocity c. Suppose it is decelerated to zero velocity while doing work on its surroundings and while no change occurs in its thermodynamic state or its elevation within the Earth's gravitational field. This capacity for doing work is its kinetic energy and is given by

$$E_k = -W = -\int_c^0 F \, ds \tag{4.1}$$

where W is the work done on the system and F is the accelerating force acting in the direction of the velocity vector c. Therefore

$$E_k = -\int_c^0 ma \, ds \tag{4.2}$$

where a is the acceleration (from Newton's second law). As the mass is constant and as acceleration is the rate of change of velocity

$$E_k = -m\int_c^0 \frac{dc}{dt} \, ds \tag{4.3}$$

$$= -m\int_c^0 \frac{ds}{dt} \, dc \tag{4.4}$$

$$= -m\int_c^0 c \, dc \tag{4.5}$$

$$= -m\left[\frac{c^2}{2}\right]_c^0. \tag{4.6}$$

Hence

$$E_k = m\frac{c^2}{2}. \tag{4.7}$$

The specific kinetic energy is obtained by dividing by the mass to give

$$e_k = \frac{c^2}{2}. \tag{4.8}$$

The right-hand side of Equation (4.8) would have the SI units [$m^2 \, s^{-2}$], which are equivalent to J/kg.

4.2.2 Gravitational potential energy

Gravitational potential energy (which is often abbreviated to just 'potential energy') is the energy possessed by a mass due to its position in the gravitational field measured from a reference datum level.

Suppose a system has a mass m and an elevation z above the reference level in a gravitational field that has an acceleration due to gravity of g. If it is lowered to the datum level without any change in its velocity or in any of its thermodynamic properties, it can do work on its surroundings. This capacity for doing work is its potential energy and is given by the negative of the work done on the system:

$$E_p = -W = -\int_z^0 F \, ds \tag{4.9}$$

where F is the force exerted on the system in the upwards direction by its surroundings (which matches the downwards force of gravity) and s is the displacement measured in the upwards direction. Hence

$$E_p = -\int_z^0 mg \, ds \tag{4.10}$$

$$= -mg \int_z^0 ds \tag{4.11}$$

$$= mgz. \tag{4.12}$$

The specific potential energy is obtained by dividing by the mass to give

$$e_p = gz. \tag{4.13}$$

The right-hand side of Equation (4.13) would have the SI units [m s^{-2} m], which are equivalent to J/kg.

4.2.3 Internal energy

Internal energy is the molecular energy possessed by a substance. It includes the kinetic energy of translation, rotation and vibration of the molecules and their constituents, and the potential energy associated with the forces between these. The potential energy associated with forces between molecules and submolecular constituents can be likened to the energy stored in a coil spring that has been compressed or extended.

Internal energy is the macroscopic term for the total of the energies possessed by a substance at the molecular and submolecular levels. Its symbols and units are shown in Table 3.1. In engineering thermodynamics, changes in internal energy, rather than absolute values, are important.

Not all of the energy possessed by a substance at the molecular or submolecular levels is convertible to work at the macroscopic level. Internal energy is distributed

Table 3.1

Property	*Symbol*	*Unit*
Internal energy	U	J
Specific internal energy	u	J/kg

over the immense number of modes of energy possession of the substance in the system.[7] A single molecule could have translational kinetic energy in the x, y and z directions. Some molecules can also have significant rotational kinetic energy about each of three orthogonal axes. Six modes of energy possession have thus been cited for just one molecule. It would not be practical to attach a mechanical device to an individual molecule to extract, as work, all the kinetic energy it possessed at a given moment in time. It would be even less practical to attach separate mechanical devices to a vast collection of molecules to provide a combined work output from all the kinetic energy they possessed.

4.2.4 Total energy

The *total energy* of a system is the sum of all the types of energy it possesses. For a system that contains only a compressible substance, the total energy is given by Equation (4.14) and the total specific energy by Equation (4.15):

$$E = E_k + E_p + U \tag{4.14}$$

$$e = e_k + e_p + u. \tag{4.15}$$

4.3 The principle of conservation of energy

The *principle of conservation of energy* states that energy cannot be created or destroyed, but only changed in form. This principle, like the principle of conservation of mass, can almost always be applied in engineering thermodynamics. Normally, the properties mass and energy are separately conserved.

The principle of conservation of energy is based on observation. As energy can take many different forms and as there are different modes of energy transfer, the set of observations on which the principle of conservation of energy depends is a very extensive one.

4.4 Temperature

The property 'temperature' can be explained as sensations of 'hotness' or 'coldness'. However, these sensations cannot provide an accurate measure of temperature. An explanation of temperature that allows precise measurements to be made is given in the following paragraphs.

Two systems are said to have the same temperature (or to be in *mutual thermal equilibrium*) if no change would occur in either system if each were enclosed by

7. This is some multiple greater than or equal to three of Avogadro's number for 1 kilogram mole of substance.

a rigid, non-adiabatic, impermeable boundary and the two systems together were isolated from the surroundings.

The *zeroth law of thermodynamics* states that if two systems each have the same temperature as a third system they have the same temperature as each other. This law is based on observation. Because this law is so fundamental to our everyday usage of the term 'temperature', the zeroth law may seem obvious and trivial. It may help, therefore, to consider a totally different situation in which this type of law would not necessarily hold. Suppose person A and person B like each other and suppose person B and person C like each other. It does not necessarily follow that person A and person C like each other.

On the basis of these concepts of equality of temperature, various properties of 'special systems' (thermometers) can be used to construct temperature scales; for example,

- the pressure of a constant volume of gas
- the volume of a liquid (measured by the length of a thin liquid column attached to a fat bulb of liquid)
- the electromotive force (e.m.f.) of a thermocouple
- the electrical resistance of a conductor.

These and other systems are used for practical temperature measurements. A *thermometer* is a special system that has a property that is sensitive to temperature. The different types of thermometers can be calibrated to a common practical temperature scale. Temperature is thus a measurable thermodynamic property of a system that is in equilibrium.

Temperature is an intensive property whose difference on a fixed scale for any two equilibrium systems determines the direction in which the spontaneous transfer of internal energy would occur between them if each were enclosed by a rigid, non-adiabatic, impermeable boundary and the two systems together were isolated from the surroundings.

It is found that certain systems have a temperature that does not vary as long as the specification of the system is maintained. The temperature of a system that contains the solid and liquid phases of a pure substance in equilibrium at a specified pressure is called the *fusion or freezing or melting temperature* of that substance at that pressure and is invariant. This temperature may also be referred to, less formally, as the *freezing point*. The temperature of a system that contains the liquid and vapour phases of a pure substance in equilibrium at a specified pressure is called the *boiling, or saturation, or evaporation or condensation temperature* of that substance at that pressure and is invariant. Less formally, this temperature may be described as the *boiling point*. The temperature of a system that contains the solid, liquid and vapour phases of a pure substance in equilibrium is called the *triple point temperature* of that substance and is invariant. The pressure of such a system is also invariant.

Such fixed-temperature systems are useful for the establishment and maintenance of a common practical temperature scale where different types of thermometers can be used. The Celsius scale was originally defined in terms of the boiling and freezing points of water at standard atmospheric pressure.

The adjective *isothermal* means 'at the same temperature'. An *isothermal process* is a process during which the temperature does not change. For instance, the pressure of a gas could be increased while its temperature is kept constant.

It will be seen in Chapter 13, section 13.7, that the second law of thermodynamics allows an absolute temperature scale to be defined in terms of the characteristics of an ideal engine. The kelvin scale is defined in this way. Such a scale is independent of the properties of particular systems used as thermometers. An absolute or thermodynamic temperature scale has a zero level that can be approached, but not attained: absolute zero temperature.

All thermometers are now calibrated by reference to an absolute scale using the triple point temperature of water as the single fixed reference point. The kelvin unit is defined as

$$1\ \mathrm{K} \triangleq \frac{\text{temperature of the triple point of water}}{273.16}. \tag{4.16}$$

The magnitude of the kelvin unit has been chosen so that it corresponds with the degree Celsius (°C) when used to measure the temperature difference between the freezing and boiling points of water at standard atmospheric pressure.

The Celsius scale is now formally defined in terms of the kelvin scale by Equation (4.17):

$$\frac{t}{[°\mathrm{C}]} \triangleq \frac{T}{[\mathrm{K}]} - 273.15 \tag{4.17}$$

where t = conventional temperature (with a zero value at the freezing point of water) (units: °C)

T = absolute temperature (units: K).

$$1\ °\mathrm{C}\ \text{temperature difference} = 1\ \mathrm{K}\ \text{temperature difference}. \tag{4.18}$$

The triple point temperature of water is 0.01 °C and occurs at a pressure of 0.000 611 MPa, whereas the freezing point of water at standard atmospheric pressure is 0 °C. The difference of 0.01 °C between these two temperatures is the reason why Equation (4.16) incorporates the constant 273.16, whereas Equation (4.17) incorporates the constant 273.15.

4.5 Heat

When two systems that have the same temperature are brought into contact, there is no net transfer of internal energy between them. However, if two systems that do not have the same temperature are brought into contact, internal energy will be redistributed within them until thermal equilibrium is established. There will be a net transfer of internal energy across their common boundary from the system at the higher temperature to the system at the lower temperature. This is heat transfer.

Heat (which can also be described as *heat transfer*) is the transfer of internal energy across a boundary owing to a temperature difference. The SI unit of heat is the joule. Heat is transitory and exists only while there is energy transfer due to a temperature difference.

An *adiabatic process* is a process in which there is no heat transfer between a system and its surroundings. An *adiabatic boundary* is a boundary that does not permit heat transfer. Such a boundary constraint is approximated in practice by the use of insulating materials (poor heat conductors).

4.5.1 Measurement of heat transfer

Heat transfer is measured indirectly, often by measuring the changes in the properties of a system that receives or provides the heat transfer. For instance, the property specific internal energy has been carefully determined for water substance over a very wide range of equilibrium states. If heat transfer is the only mechanism by which energy enters or leaves a closed system that contains water substance, the amount of the heat transfer can be determined from the mass of the water and the change in its specific internal energy between the initial and final equilibrium states. The measurement of the amount of the heat transfer could thus be based on the measurement of the mass of the water, the volume of the system and the temperatures at the initial and final equilibrium states of the water. The first law of thermodynamics, which is described in Chapter 10, formally describes the equivalence of heat and work in so far as they are both modes of energy transfer and allows for the formal definition of a difference in internal energy between states.[8]

4.6 The sign convention for heat

When the focus of attention is on just one system it is convenient to define a positive and a negative direction for heat transfer in relation to that system. *Heat transfer to a system* is taken to be *positive* with respect to the system. *Heat transfer from a system* is taken to be *negative*. As in the case of the sign convention for work, this convention can be overridden by explicitly indicating the positive direction for heat transfer. Whenever a heat transfer quantity appears within a thermodynamic relation without qualification, the convention will be taken to apply.

Heat and work are interactions that involve energy transfer between systems. The

8. Historically, the science of measuring heat transfer, known as *calorimetry*, existed before it was realized that heat was a form of energy transfer: heat was regarded as an invisible fluid called *caloric* that could flow from one system to another. A commonly used unit, the calorie, was the amount of heat transfer necessary to raise the temperature of 1 gram of liquid water through 1 degree Celsius. This unit, which is now defined more precisely, is still sometimes used, but is not an SI unit.

sign conventions for heat and for work can be generalized to include all energy transfers across the boundary of a system. *Energy transfer into a system* is taken to be *positive* with respect to the system. *Energy transfer out of a system* is taken to be *negative* with respect to the system.

In diagrams in this textbook a special arrowhead (—►) is used to identify heat transfer or a heat transfer rate. This is a solid triangular arrowhead. See, for instance, Figure 11.1 of Chapter 11 and Figure 12.7 of Chapter 12. If the arrow is labelled with a symbol, then the direction of the arrow is to be taken as the positive direction for heat transfer represented by that symbol. Thus, the combination of a heat transfer arrow and a symbol takes precedence over the normal convention that heat transfer into a system is positive. This is also the preferred method of dealing with the algebraic sign of heat transfer when the focus of attention is on more than one system. Where the direction of an energy transfer as heat is known, it is usual to draw the arrow from the system that provides the heat transfer to the system that receives the heat transfer, i.e. from the system at the higher temperature to the system at the lower temperature.

4.7 Net changes and effects

A *net change* is the change that remains after all changes have occurred. For instance, if the temperature of a system changes from 10 °C to 25 °C and then to 17 °C, the net temperature change is from 10 °C to 17 °C. In everyday language and in the terminology of thermodynamics, an *effect* can be either the bringing about of a change or the result of this, which is the change. Work and heat transfer are both effects (in the 'bringing about' sense) that cause the energy of a system to change. The transport of an extensive property across a boundary is an effect. The transfer of a particular chemical species across a boundary is another example. A *net effect* is the combined effect of all effects. If a system receives various amounts of energy as heat transfer and rejects various amounts of energy as heat transfer, then the net heat transfer effect (with the usual sign convention) is the algebraic sum of all positive and negative heat transfer quantities. A net work effect, a net transport of internal energy or a net transfer of oxygen could be similarly evaluated.

4.8 Summary

The principles of conservation of mass and energy have been introduced. Definitions and explanations have been provided for the concepts energy, temperature and heat. Kinetic energy, gravitational potential energy, internal energy and total energy have been explained. The sign convention for heat has been presented, as has the general sign convention for energy transfer across the boundary of a system. Net changes and effects have been defined and examples given.

The student should be able to

- *define*
  — energy, kinetic energy, gravitational potential energy, internal energy, total energy, when two systems have the same temperature, when two systems are in mutual thermal equilibrium, a thermometer, temperature, fusion temperature, freezing temperature, melting temperature, freezing point, boiling temperature, saturation temperature, evaporation temperature, condensation temperature, boiling point, triple point temperature, isothermal, an isothermal process, the kelvin unit, heat, heat transfer, an adiabatic process, an adiabatic boundary, positive heat transfer, negative heat transfer, positive energy transfer, negative energy transfer, a net change, an effect, a net effect

- *explain*
  — the basis of a thermometer and of practical temperature scales, how heat transfer can be measured, the relationship between conventional temperature and absolute temperature

- *state*
  — the principle of conservation of mass, the principle of conservation of energy, the zeroth law of thermodynamics, the sign convention for heat, the sign convention for energy transfer.

4.9 Self-assessment questions

4.1 What is the specific kinetic energy of a fluid that has a velocity of 10 m/s? What is the specific potential energy of a system that is at a height of 10 m above a specified datum level for zero potential energy in the gravitational field? Take the acceleration due to gravity as 9.81 m/s^2.

4.2 An unmarked mercury-in-glass thermometer was calibrated as follows:
(a) It was immersed in melting ice and the position of the top of the mercury column was marked on the stem and assigned the value 0 °C.
(b) It was immersed in boiling water and the new position of the top of the mercury column was marked. This was assigned the value 100 °C.
The distance between the two marks was found to be 220 mm. Assuming a linear relationship between temperature and column length, what would be the measured temperature when the top of the mercury column was 49 mm above the mark for melting ice?

CHAPTER 5

Mechanical work processes of closed systems

Positive work at the boundary of a system is a mechanism of energy transfer into the system and tends to increase its energy. Negative work at the boundary of a system is a mechanism of energy transfer out of the system and tends to decrease its energy. This chapter provides more information on the evaluation of mechanical work for closed systems.

5.1 Normal or displacement work processes of closed systems

5.1.1 *Equilibrium normal or displacement work*

The equilibrium normal or displacement work for a process of a closed system that contains a compressible fluid is given by the following equations:

$$W = -\int p \, dV \qquad (3.16, \text{Chapter 3})$$

$$= -(\text{area below } p\text{–}V \text{ process curve}). \qquad (3.18, \text{Chapter 3})$$

The work can be evaluated if the path can be described, i.e. provided the pressure is known as a function of the volume for the entire process. There are various special cases where the equilibrium process path is described by a relatively simple analytical expression. Some examples are given below.

Constant pressure process

$$W = -\int_1^2 p \, dV$$

$$= -p\int_1^2 dV = -p(V_2 - V_1) = -p\Delta V. \qquad (5.1)$$

Constant volume process

$$W = 0. \qquad (5.2)$$

Hyperbolic process

A *hyperbola* is any curve that has the form xy = const. A *hyperbolic process* is a process that can be described by an equation in terms of absolute pressure and volume that has the form of a hyperbola; namely, pV = const. It is found experimentally that certain real processes can be represented accurately in this way. The expansion of a gas at constant temperature can usually be described as a hyperbolic process. In such a process heat transfer must occur to the gas in order to maintain its temperature constant.

$$p_1 V_1 = p_2 V_2 = pV = \text{const} \tag{5.3}$$

$$\therefore p = \frac{p_1 V_1}{V}$$

$$W = -\int_1^2 p \, \mathrm{d}V$$

$$= -\int_1^2 \frac{p_1 V_1}{V} \, \mathrm{d}V = -p_1 V_1 (\ln V_2 - \ln V_1) = -p_1 V_1 \ln \frac{V_2}{V_1}.$$

Hence

$$W = p_1 V_1 \ln \frac{V_1}{V_2}. \quad \text{(This is the hyperbolic work.)} \tag{5.4}$$

Also, since

$$\frac{p_2}{p_1} = \frac{V_1}{V_2}$$

$$W = p_1 V_1 \ln \frac{p_2}{p_1}. \tag{5.5}$$

Polytropic process

A *polytropic process* is a process that can be described by an equation that has the form pV^n = const, where p is absolute pressure. It is found experimentally that certain real processes can be represented accurately in this way. The expansion of a gas can often be described as a polytropic process. The extent and direction of the heat transfer influences the value of the polytropic exponent (or index) n.

$$p_1 V_1^n = p_2 V_2^n = pV^n = \text{const} \tag{5.6}$$

$$\therefore p = \frac{\text{const}}{V^n}$$

$$W = -\int_1^2 p \, dV$$

$$= -\int_1^2 \frac{\text{const}}{V^n} \, dV$$

$$= -\text{const} \int_1^2 \frac{dV}{V^n}$$

$$= -\text{const} \left[\frac{V^{-n+1}}{-n+1} \right]_1^2$$

$$= -\text{const} \left[\frac{V_2^{-n+1} - V_1^{-n+1}}{-n+1} \right]$$

$$= -\frac{p_2 V_2^n V_2^{-n+1} - p_1 V_1^n V_1^{-n+1}}{-n+1}$$

$$= -\frac{p_2 V_2 - p_1 V_1}{-n+1}.$$

Hence

$$W = \frac{p_2 V_2 - p_1 V_1}{n-1}. \quad \text{(This is the polytropic work.)} \qquad (5.7)$$

EXAMPLE 5.1

Case study: an air standard Otto cycle

This case study is an analysis of a simplified and somewhat idealized form of the cycle of an internal combustion engine. The cycle is modelled as a series of processes undergone by a closed system containing air. It is described in terms of pressure and volume changes.

Air at a pressure of 0.1 MPa is enclosed by a cylinder and piston. Owing to the movement of the piston its volume is reduced from 1 litre to 1/8 litre in a polytropic process with an exponent of 1.4. The pressure is then increased at constant volume to 4.8 MPa. The air is brought back to its original volume in another polytropic process with the same exponent. Finally it is brought to the initial pressure in a constant volume process. This completes one cycle. Sketch the cycle on a p–V diagram. Calculate the work for each process and calculate the average power output if there are 3000 cycles per minute.

SOLUTION

$$p_1 V_1^{1.4} = 0.1 \times 10^6 \text{ [N m}^{-2}] \, (1 \times 10^{-3})^{1.4} \text{ [m}^3]^{1.4} = 6.310 \text{ [N m}^{-2}][\text{m}^3]^{1.4}.$$

For process 1 → 2

$$pV^{1.4} = p_1 V_1^{1.4} = 6.310 \text{ [N m}^{-2}][\text{m}^3]^{1.4};$$

at $V = 0.5 \times 10^{-3} \text{ m}^3$

$$p = \frac{6.310 \text{ [N m}^{-2}][\text{m}^3]^{1.4}}{(0.5 \times 10^{-3})^{1.4} \text{ [m}^3]^{1.4}} = 0.264 \times 10^6 \text{ N/m}^2,$$

at $V = 0.25 \times 10^{-3} \text{ m}^3$

$$p = \frac{6.310}{(0.25 \times 10^{-3})^{1.4}} \text{ N/m}^2 = 0.696 \times 10^6 \text{ N/m}^2,$$

at $V_2 = 0.125 \times 10^{-3} \text{ m}^3$

$$p_2 = \frac{6.310}{(0.125 \times 10^{-3})^{1.4}} \text{ N/m}^2 = 1.838 \times 10^6 \text{ N/m}^2.$$

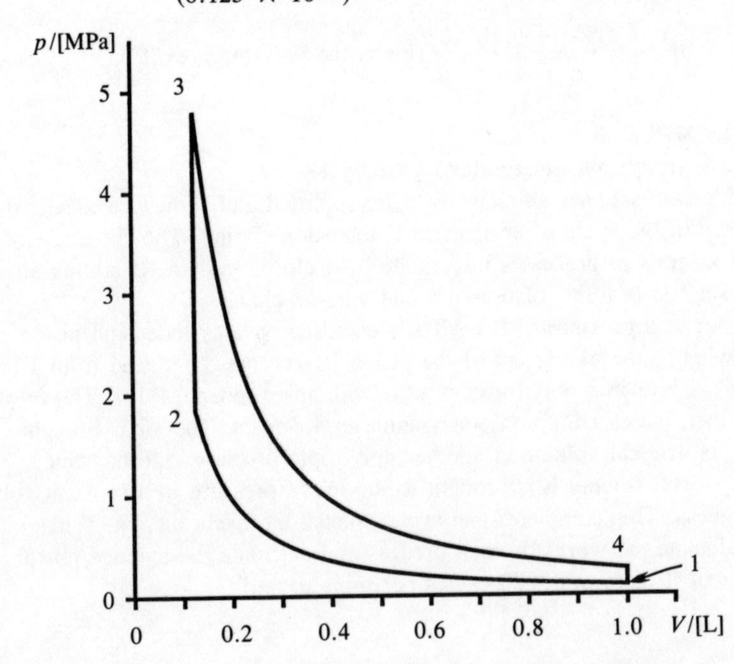

Figure 5.1 *Sketch of the cycle for Example 5.1 on a p–V diagram.*

Curve 1 → 2 in Figure 5.1 is sketched from the above information.

For process 3 → 4

$$pV^{1.4} = (4.8 \times 10^6)\,[\text{N m}^{-2}]\,(0.125 \times 10^{-3})^{1.4}\,[\text{m}^3]^{1.4}$$

$$= 16.48\,[\text{N m}^{-2}][\text{m}^3]^{1.4};$$

at $V = 0.25 \times 10^{-3}\ \text{m}^3$

$$p = \frac{16.48}{(0.25 \times 10^{-3})^{1.4}}\ \text{N/m}^2 = 1.819 \times 10^6\ \text{N/m}^2,$$

at $V = 0.5 \times 10^{-3}\ \text{m}^3$

$$p = \frac{16.48}{(0.5 \times 10^{-3})^{1.4}}\ \text{N/m}^2 = 0.689 \times 10^6\ \text{N/m}^2,$$

at $V_4 = 1.0 \times 10^{-3}\ \text{m}^3$

$$p_4 = \frac{16.48\,[\text{N m}^{-2}]}{(1 \times 10^{-3})^{1.4}} = 0.261 \times 10^6\ \text{N/m}^2.$$

Curve 3 → 4 in Figure 5.1 is sketched from the above information.

For process 1 → 2

$$W = \frac{p_2V_2 - p_1V_1}{n - 1}$$

$$W = \left(\frac{(1.838)(0.125) - (0.1)(1)}{0.4}\right) \times 10^6 \times 10^{-3}\,[\text{N m}^{-2}][\text{m}^3]$$

Answer

$$= 324.4\ \text{J}.$$

For process 2 → 3

Answer

$$W = 0\ \text{J}.$$

For process 3 → 4

$$W = \frac{p_4 V_4 - p_3 V_3}{n - 1}$$

$$W = \left(\frac{(0.261)(1) - (4.8)(0.125)}{0.4}\right) \times 10^3 \text{ J}$$

Answer

$$= -847.5 \text{ J.}$$

For process 4 → 1

Answer

$$W = 0 \text{ J.}$$

For all four processes

$$\Sigma W = 324.4 + 0 - 847.5 + 0 = -523.1 \text{ J}$$

$$\dot{W}_{avg} = -523.1 \text{ [J]} \frac{3000}{60} \text{ [s}^{-1}\text{]}$$

$$= -26\,155 \text{ W}$$

$$= -26.16 \text{ kW.}$$

The negative sign of ΣW and of $\dot{W}_{avg}$ indicates that the cycle produces a net work *output* and an average power *output*.

The average power output is given by

Answer

$$\dot{W}_{avg,out} = -\dot{W}_{avg} = 26.16 \text{ kW.}$$

5.1.2 Non-equilibrium normal or displacement work

For a closed system that contains only a compressible fluid, the work cannot be determined by evaluating the negative integral of the pressure with respect to the volume (Equation (3.16) of Chapter 3) if the system is not in mechanical equilibrium during the process. This is because a single pressure value cannot represent the normal force per unit area at every point on the boundary.

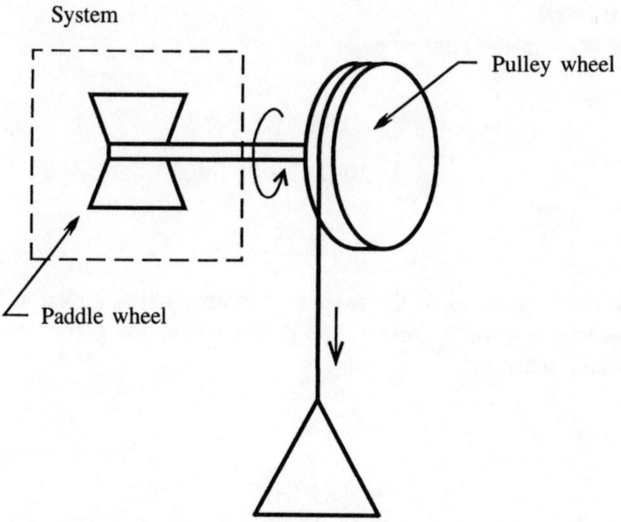

System

Pulley wheel

Paddle wheel

Figure 5.2 *Shaft stirring work on a closed system that contains only a compressible fluid.*

5.2 Shear or shaft work processes of closed systems

In general, the energy of a system can be increased by positive shear (or shaft) work or decreased by negative shear (or shaft) work. An example of this situation would be a closed system that contained a clockwork mechanism. Shaft work could be done on or by this system.

If a system contains only a compressible fluid in equilibrium, neither positive nor negative *equilibrium* shaft work is possible. Positive non-equilibrium shaft work is possible; for instance, stirring work,[9] as shown in Figure 5.2. A system always undergoes a non-equilibrium process when stirring work is done on it. Because of fluid friction there would be shear forces within the fluid while the stirring process continued. These forces would change as the fluid came to rest if the system were isolated. Once a system is in equilibrium it cannot cause a paddle wheel to rotate and negative shaft work cannot occur.

EXAMPLE 5.2

Consider a situation of the type illustrated in Figure 5.2. Suppose a mass of 10 kg descends at a steady rate of 2 m/s while doing work on the system. Take the acceleration due to gravity to be 9.81 m/s^2. What is the rate at which stirring work is done on the system?

9. Stirring work is defined formally in Chapter 10, section 10.4.

SOLUTION

The tension in the cord is given by

$$F = mg$$

$$= 10\ [\mathrm{kg}]\ 9.81\ [\mathrm{m\ s^{-2}}]$$

$$= 98.1\ \mathrm{N}.$$

Let the rate of travel of the mass in the downwards direction be $\dot{y}$. The rate at which work is done on the pulley wheel and thence on the system as stirring work is

$$\dot{W} = F\dot{y}$$

$$= 98.1\ [\mathrm{N}]\ 2\ [\mathrm{m\ s^{-1}}]$$

Answer

$$= 196.2\ \mathrm{W}.$$

5.3 Summary

Details and examples have been given of how work is evaluated for closed systems.

The student should be able to

- *define*
  — a hyperbola, a hyperbolic process, a polytropic process
- *state and derive*
  — expressions for the equilibrium normal work for a constant pressure process, a constant volume process, a hyperbolic process and a polytropic process
- *calculate*
  — normal and shear work and work rates for closed systems.

5.4 Self-assessment questions

5.1 Figure 5.3 shows an unloaded gas spring consisting of a cylinder with a frictionless leak-tight piston. The compression process is known to be polytropic, with an exponent of 1.3.

(a) Determine the ideal work done on the system if the piston is moved 80 mm into the cylinder.

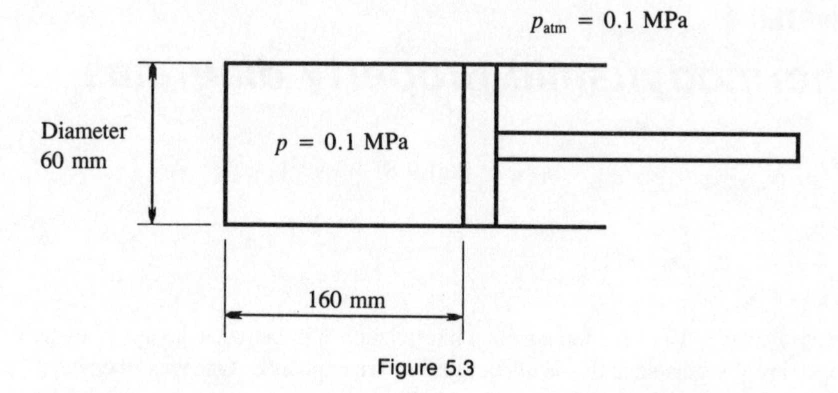

Figure 5.3

(b) Determine the work done by an external agency that exerts a force on the piston rod.

5.2 A fixed mass of gas is contained in a cylinder by a frictionless leak-tight piston. It occupies a volume of 0.1 m^3 at a pressure of 0.4 MPa absolute. The gas is first compressed in a hyperbolic process to a volume of 0.015 m^3 and its volume is then increased at constant pressure to 0.05 m^3. Sketch the processes on a $p-V$ diagram. Calculate the work for each and the net work.

5.3 A closed system containing a gas has an initial volume of 9.373 L and an initial pressure of 0.120 MPa. The gas is compressed hyperbolically until the volume is 1.338 L. It is then compressed polytropically with an index of 1.667 until its pressure reaches 4.340 MPa. From this pressure it is expanded hyperbolically until its volume reaches 3.502 L. From this state it is expanded polytropically to the initial state.

(a) Sketch the complete cycle, which comprises four processes, on a $p-V$ diagram.
(b) Calculate the polytropic exponent for the last process.
(c) Determine the work for each process.
(d) Determine the net work for the cycle.

CHAPTER 6

Thermodynamic property diagrams

In this chapter some of the fundamental interrelationships between the thermodynamic properties of a substance that is in equilibrium are explained. One way of developing an intuitive understanding of these interrelationships is to draw graphs of them. Some of the most useful graphical representations are presented.

6.1 *p–v–T* equilibrium diagrams

Consider an equilibrium process in which heat transfer occurs to a system that contains water substance at constant pressure as shown in Figure 6.1.

At a given pressure, the temperature at the various states, A to E, can be plotted on a p–T state diagram, as shown in Figure 6.2. Similarly, the pressure and the specific volume can be plotted on a p–v state diagram, as shown in Figure 6.3. The specific volume is the volume divided by the mass present, which is fixed. The experiment can be repeated at different pressures. Hence, the equilibrium lines that

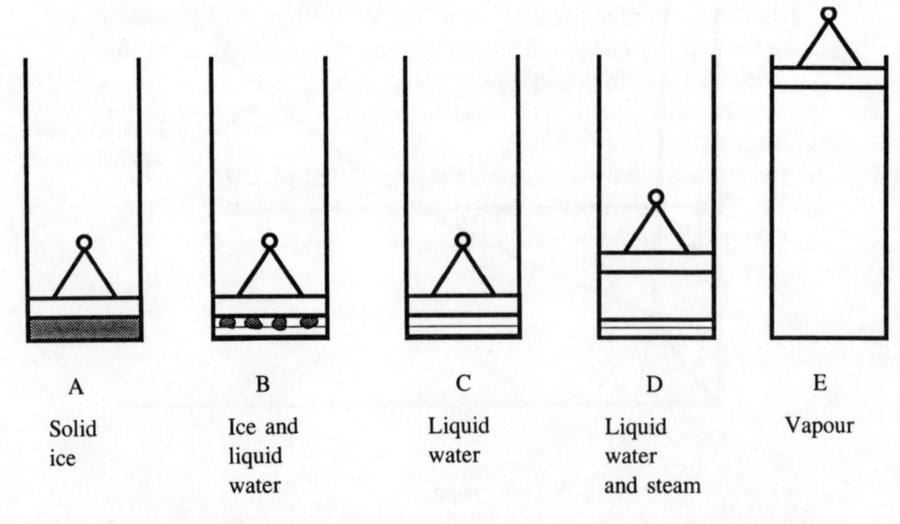

Figure 6.1 *A hypothetical experiment in which heat transfer occurs to a system at constant pressure and the pure substance within it, water, undergoes phase changes.*

mark phase transitions can be plotted as shown in Figures 6.2 and 6.3. An *equilibrium phase diagram* is any diagram that indicates the phases of various equilibrium states of a substance. Figures 6.2 to 6.7 are all equilibrium phase diagrams.

A notable feature of Figure 6.2 is that at very low pressures water substance can pass from the solid phase to the vapour phase at constant pressure without going through the liquid phase. A direct phase change from vapour to solid is also possible at these very low pressures. Such phase changes from solid to vapour or from vapour to solid are described as *sublimation*.

Water is different to most other substances in that liquid water expands on freezing and solid ice contracts on melting. The part of the constant pressure process of Figure 6.1 that includes melting is illustrated in the enlarged inset of Figure 6.3. As the state changes from A to C the path is A, B_i, B, B_f, C.

All the data represented in Figures 6.2 and 6.3 can be combined to yield a three-dimensional surface that represents the interrelationships between the properties p, v and T, at equilibrium, for water substance. This is shown in Figure 6.4. The $p-T$ diagram in Figure 6.2 and the $p-v$ diagram in Figure 6.3 are two different orthogonal projections of the $p-v-T$ surface.

Figure 6.5 is a repetition of the $p-v$ diagram of Figure 6.3, but shows only the liquid and vapour phases. Some lines of constant temperature are also shown on this diagram. The diagram, like the others shown in this chapter, is not drawn to scale.

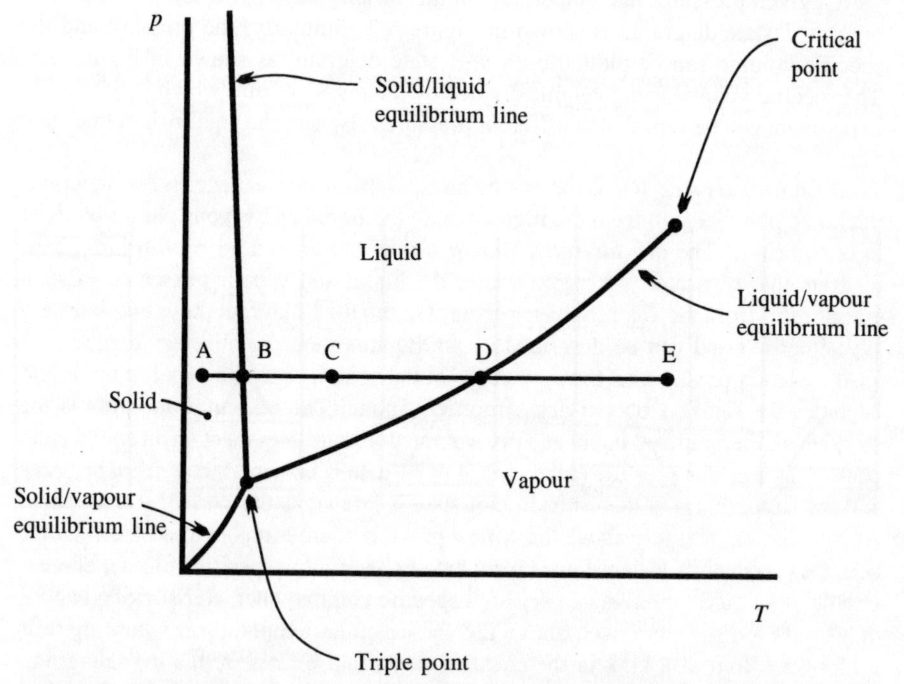

Figure 6.2 *Pressure versus temperature equilibrium phase diagram for water substance. States* A *to* E *refer to the constant pressure experiment illustrated in Figure 6.1.*

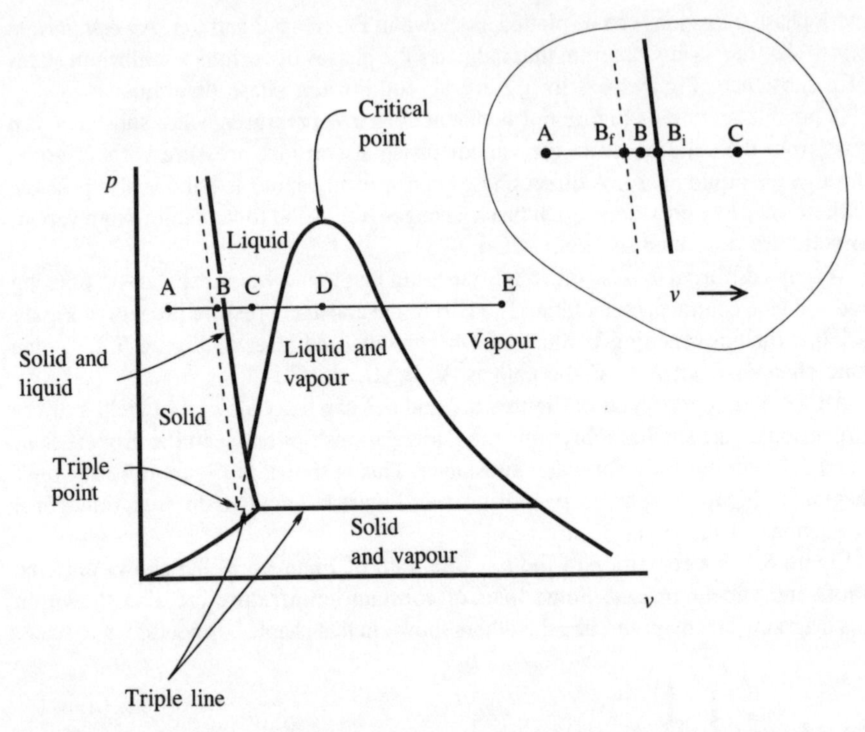

Figure 6.3 *Pressure versus specific volume equilibrium phase diagram for water substance. States* A *to* E *refer to the constant pressure experiment illustrated in Figure 6.1.*

The *saturated liquid line* is the line on an equilibrium phase diagram that separates the liquid phase region from the region where the liquid and vapour phases co-exist in equilibrium. The *dry saturated vapour line* is the line on an equilibrium phase diagram that separates the region where the liquid and vapour phases co-exist in equilibrium from the vapour phase region. On the $p-T$ diagram these two lines are superimposed and can be described as the liquid/vapour equilibrium line.

At a certain pressure on the $p-v$ diagram the specific volume of saturated liquid becomes the same as that of dry saturated vapour. The *critical point state* is the state where the saturated liquid and dry saturated vapour lines meet on an equilibrium phase diagram. The pressure at the critical point state is known as the *critical pressure* and the temperature at the critical point state is known as the *critical temperature*. At any constant pressure above the critical pressure there is no obvious phase change boundary as water substance goes from having a very low specific volume, characteristic of a liquid, to having a very high specific volume, characteristic of a vapour or gas. At temperatures well above the critical point temperature, where there is no longer a hint of a kink in the constant temperature lines on the $p-v$ diagram, Figure 6.5, the substance can be described as a gas. A substance can also be described as a gas at pressures that are very low compared with the saturation pressure at a

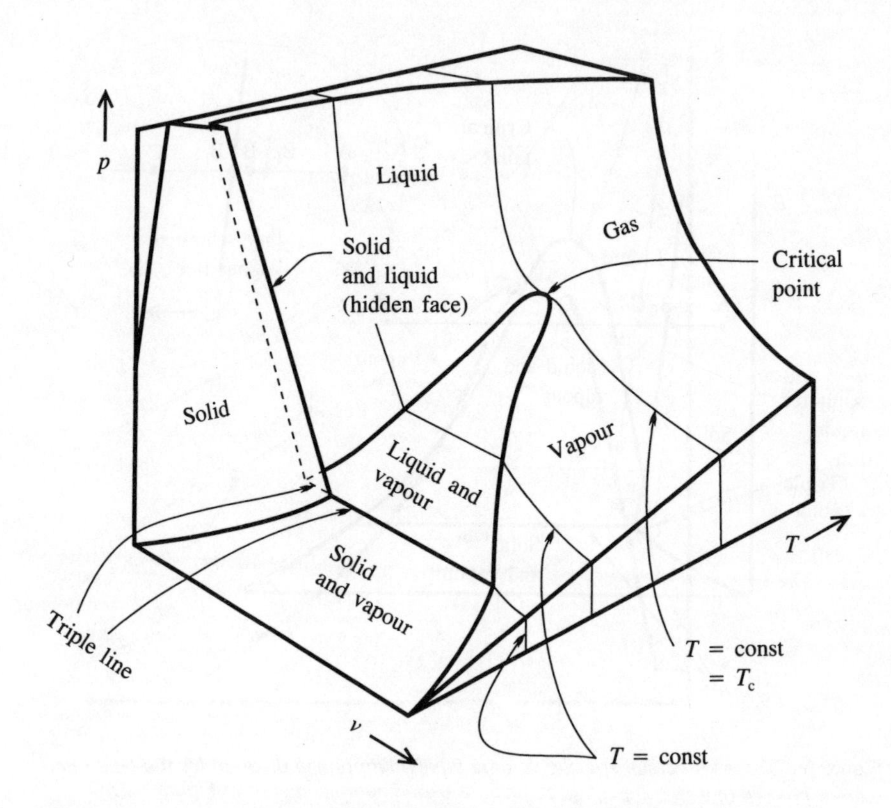

Figure 6.4 *A three-dimensional surface representing the interrelationships between the three properties p, v and T of water substance in the solid, liquid and vapour phases.*

given temperature. A gas differs from a vapour only in that it can be described by a simpler equation of state.

At low pressures the specific volume of saturated liquid is minute in comparison with that of dry saturated vapour. Also, the saturation pressure of a mixture of liquid and vapour at temperatures near the triple point temperature is minute compared with the pressure of a saturated mixture near the critical point. For this reason, linear pressure and specific volume axes are not suitable for representing the range of magnitudes involved. Equilibrium phase diagrams similar to those shown in Figure 6.3 and Figure 6.4 are obtained by using logarithmic scales for the pressure and specific volume axes.

6.2 Saturation properties

Any state where a liquid is in equilibrium with its vapour is known as a *saturation state*. The area underneath the 'saturated liquid line' and the 'dry saturated vapour

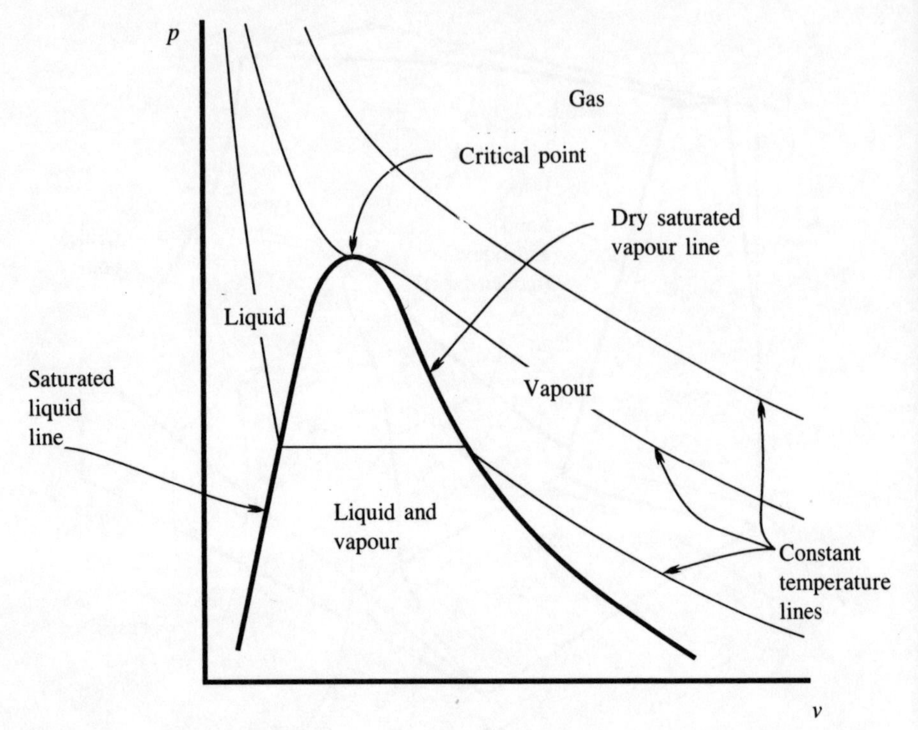

Figure 6.5 *Pressure versus specific volume equilibrium phase diagram for the liquid and vapour phases of water substance.*

line' in Figure 6.5 is known as the saturation region. In this textbook and in the steam tables that are included in Appendix A, the standard subscripts shown in Table 6.1 are used to identify saturation properties. For example,

p_s = saturation pressure at a given temperature,
h_g = specific enthalpy of dry saturated vapour at a given temperature or pressure,
$h_{fg} = h_g - h_f$ at a given temperature or pressure.

Table 6.1

f	property of saturated liquid
g	property of dry saturated vapour
fg	liquid–vapour saturation property difference
s	saturation property, *T* or *p*
c	critical point property

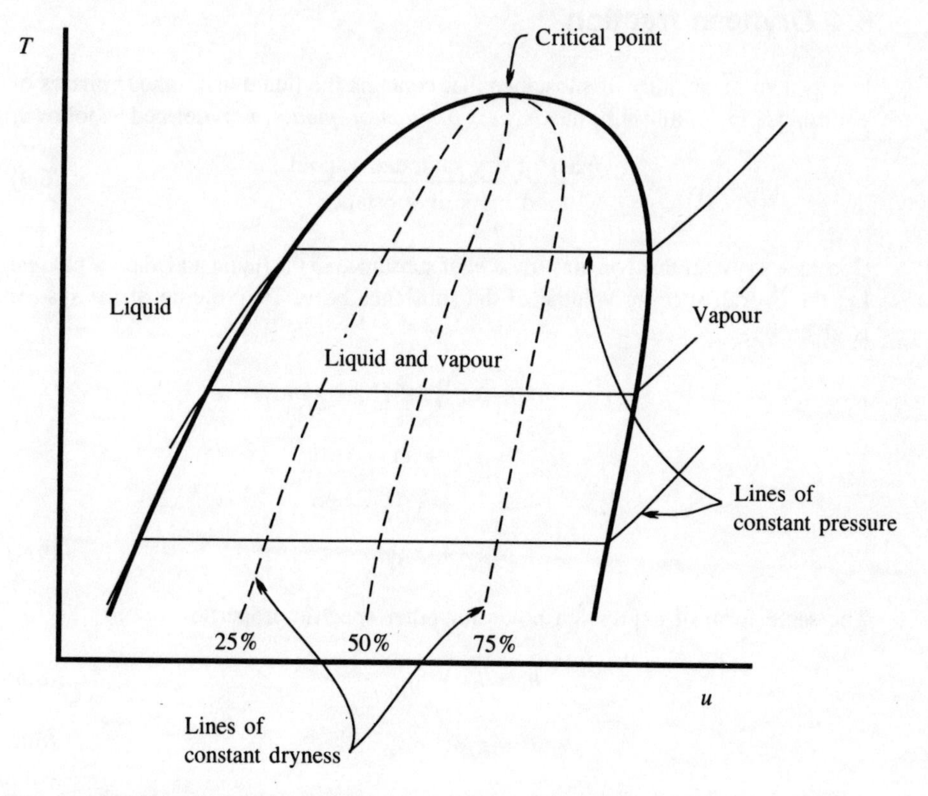

Figure 6.6 *Temperature versus specific internal energy equilibrium diagram for the liquid and vapour phases of water substance.*

6.3 Enthalpy

The combination of properties $U + pV$ frequently arises in engineering thermodynamics, especially in relation to open systems (see Chapter 11, section 11.2). It is given the name *enthalpy*. It has the units of energy and, for the moment, can be regarded as a second thermodynamic energy property.

$$H = U + pV, \tag{6.1}$$

or, in terms of intensive properties, the specific enthalpy can be written as

$$h = u + pv. \tag{6.2}$$

6.4 Dryness fraction

In a system or quantity of substance that contains the liquid and vapour phases of a substance in equilibrium, the *dryness fraction* or *quality*, x, is defined as follows:

$$x = \frac{\text{mass of dry saturated vapour}}{\text{total mass of substance}}. \tag{6.3}$$

Consider a system that contains mass m of substance in the liquid and vapour phases. Let the overall specific volume of the substance be v. The volume of the system is given by

$$V = mv = xmv_g + (1 - x)mv_f$$

$$\begin{aligned} \therefore\ v &= xv_g + (1 - x)v_f \\ &= v_f + x(v_g - v_f) \\ &= v_f + xv_{fg}. \end{aligned} \tag{6.4}$$

The same form of expression holds for other specific properties:

$$h = h_f + xh_{fg} \tag{6.5}$$

$$u = u_f + xu_{fg}. \tag{6.6}$$

In many practical situations $v_f \ll v_g$. For example, for water at 100 °C

$$v_f = 0.001\,044 \text{ m}^3/\text{kg}$$

$$v_{fg} = 1.671\,953 \text{ m}^3/\text{kg}.$$

Therefore, unless x is less than about 10% or the saturation temperature is greater than about 100 °C, Equation (6.7) can be used to calculate the specific volume of a mixture of liquid water and water vapour in equilibrium. If in doubt, however, use Equation (6.4).

$$v \approx xv_{fg} \approx xv_g. \tag{6.7}$$

The steam that leaves a boiler is likely to entrain fine droplets of saturated liquid water thrown up from the agitated surface of the boiling liquid. Such steam, which contains suspended particles of liquid water, is known as *wet steam*. It might, for instance, have a quality of 98%.

6.5 Internal energy and enthalpy diagrams

For a pure substance, such as water, a $T\text{–}u$ diagram for the liquid and vapour phases has the form shown in Figure 6.6. Lines of constant pressure and lines of constant dryness are shown on this diagram. The constant pressure lines in the liquid region are very close to the saturated liquid line: if their positions were shown to scale they would hardly be distinguishable from the saturation line. The reason for this is the low compressibility of liquids. No liquid is perfectly incompressible but, if it were, its internal energy would depend only on its temperature and not on its pressure. When a pressure is applied to an incompressible fluid, no volume change occurs and (from Equation (3.16), Chapter 3) no work is done on it. It can therefore be shown from the principle of conservation of energy (specifically, Equation (10.11), Chapter 10) that the internal energy of an incompressible fluid is not affected by a pressure change. Therefore, the internal energy of an incompressible fluid depends only on its temperature. Also, the temperature of an incompressible fluid depends only on its internal energy.

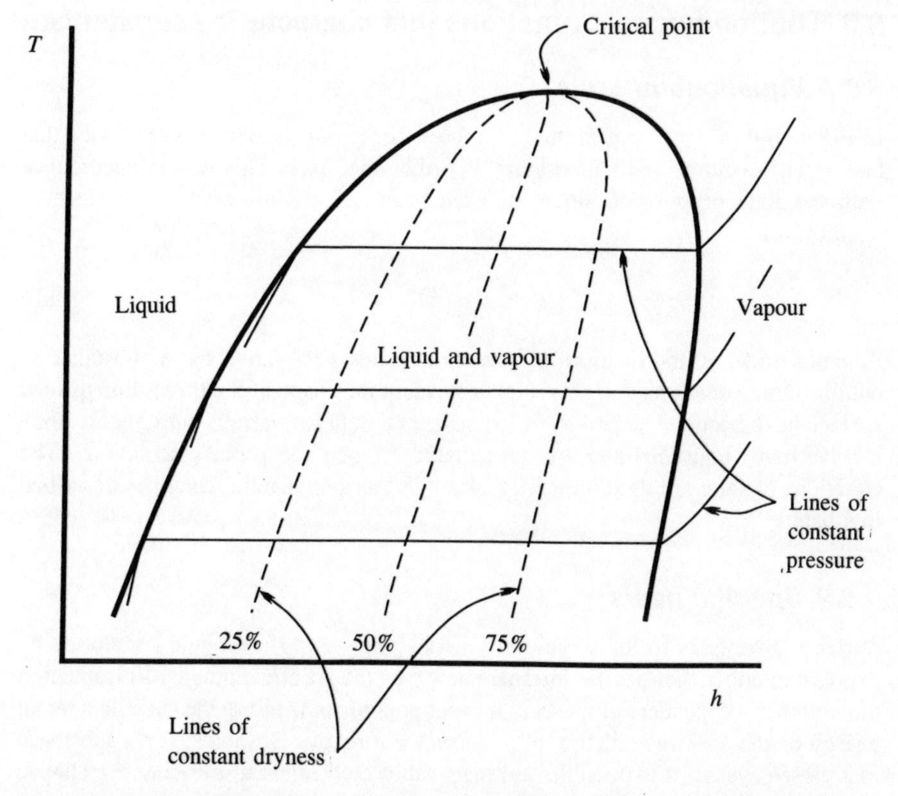

Figure 6.7 *Temperature versus specific enthalpy equilibrium diagram for the liquid and vapour phases of a pure substance.*

The liquid and dry saturated vapour saturation lines of a T–h diagram have a similar form to those of the T–u diagram. It should be noted, however, that the constant pressure lines in the compressed liquid region are to the right of the saturated liquid line, as shown in Figure 6.7, because of the effect of the pv term. In reality, the constant pressure lines in the compressed liquid region are very close to the saturated liquid line and would hardly be distinguishable from it if they were plotted to scale.

6.5.1 Latent heat

Latent heat: this term refers to the enthalpy of evaporation, h_{fg}:

$$h_{fg} = h_g - h_f. \tag{6.8}$$

It can be shown (from the steady flow energy equation, Chapter 11, and specifically Equation (11.11)) that this is also the amount of heat transfer necessary to evaporate a unit mass of liquid at constant pressure.

6.6 Thermodynamic functions that characterize substances

6.6.1 Equations of state

In Figure 6.4 pressure was plotted as a three-dimensional surface over a plane that had specific volume and temperature as orthogonal axes. This was in accordance with the state proposition since for each phase of a compressible substance in equilibrium

$$p = f(v, T). \tag{6.9}$$

Equation (6.9) is an equation of state that relates p, v and T for a substance in equilibrium. Equations that accurately represent the shape of the three-dimensional surface have been developed to fit experimental data for various substances. They involve many constants and are not suitable for pen and paper calculations. The equations of state are used indirectly, through thermodynamic tables, as described in Chapter 7.

6.6.2 Specific heats

Surfaces analogous to the three-dimensional p–v–T surface could be plotted by expressing either the specific internal energy or the specific enthalpy as a function of two other independent properties. It is not possible to measure the specific internal energy or the specific enthalpy of a substance directly. However, for a substance in a closed system it is possible and reasonably convenient to measure the change in specific internal energy for a given temperature change at constant volume. Likewise, for a stream of fluid passing through an open system, the change in specific enthalpy at constant pressure can be measured. In relation to these experimental measurements, it is therefore convenient to consider the specific internal energy and the specific enthalpy as functions of (T, v) and (T, p) respectively, as in Equations (6.10) and (6.11):

$$u = f'(T, v) \tag{6.10}$$

$$h = f''(T, p) \tag{6.11}$$

where f′ and f″ are different continuous functions for a given phase of the substance.

The *principal specific heat capacities*, or specific heats, c_v and c_p, are the slopes defined by Equations (6.12) and (6.13) of the three-dimensional surfaces defined by Equations (6.10) and (6.11) at a particular equilibrium state.

Specific heat capacity at constant volume:

$$c_v = \left.\frac{\mathrm{d}u}{\mathrm{d}T}\right|_{v=\text{const}}. \tag{6.12}$$

Specific heat capacity at constant pressure:

$$c_p = \left.\frac{\mathrm{d}h}{\mathrm{d}T}\right|_{p=\text{const}}. \tag{6.13}$$

Other specific heats can be defined, but these are the ones in common use. As the specific heat values are functions of thermodynamic properties, they are thermodynamic properties themselves.

6.6.3 Thermodynamic property tables

The data that are presented in engineering thermodynamic property tables, such as the steam tables, are derived from extensive and precise experiments. From these experiments the necessary constants for an equation of state of each particular substance and the constants for equations that describe its specific heat values are determined. Specific enthalpy values and specific internal energy values can be derived, by integration, from the specific heat data.

6.7 Summary

Some important interrelationships between the thermodynamic properties of a substance in equilibrium, such as water substance, have been explained. Thermodynamic equilibrium diagrams for the solid, liquid and vapour phases of a substance have been presented. The property enthalpy and the specific heats at constant volume and at constant pressure have been defined.

The student should be able to

- *describe and sketch*
  — a hypothetical experiment in which water substance is brought through the solid, liquid and vapour phases at various fixed pressures in order to investigate the interrelationships between p, v and T at equilibrium
- *sketch and explain*
  — the $p-T$ equilibrium diagram for the three phases of water; the $p-v$, $T-u$ and $T-h$ equilibrium diagrams for the liquid and vapour phases
- *define*
  — an equilibrium phase diagram, the saturated liquid line, the dry saturated vapour line, the critical point state, a saturation state, enthalpy, specific enthalpy, the dryness fraction, latent heat, specific heat at constant volume, specific heat at constant pressure
- *explain*
  — the $p-v-T$ equilibrium surface for water substance in three phases; the $p-v$ equilibrium diagram for the solid, liquid and vapour phases; the distinction between a gas and a vapour; sublimation; the critical pressure; the critical temperature.

6.8 Self-assessment question

6.1 With reference to Figure 6.2, explain why a layer of liquid water forms underneath the blade of an ice skate.

CHAPTER 7

The steam tables

Great amounts of data are required to describe the interrelationships between the properties of a substance such as water over the full range of property values met in engineering situations. Such data can readily be stored on a computer, either as tables or as equations of state with defining constants. For pen and paper calculations, engineers still commonly make use of published tables of thermodynamic properties. The steam tables are the most common example. A set of steam tables is included in Appendix A. This chapter describes the content of these tables and explains how to use them. Although the printed tables are increasingly being replaced by software, it will be found that an understanding of the steam tables will ensure that such software can be used properly.

7.1 Structure of the steam tables

Three tables are included in Appendix A. Table A1 gives the properties of saturated liquid water and of dry saturated steam as a function of the saturation temperature or pressure. For convenience, regular steps of both temperature and pressure have been included. Pressure or temperature values in bold type are exact to the number of decimal places shown. Table A2 is for the properties of superheated steam and includes the saturation properties for dry saturated steam in the first data column. Each row in the table is for a stated pressure. Exactly the same pressures are included as in Table A1. Therefore the superheat data are tabulated in regular increments of pressure and also in regular increments of the corresponding saturation temperature. The columns are for temperatures at regular intervals. Table A3 gives the properties of water substance at supercritical pressures, from the critical pressure up to 100 MPa. The supercritical table serves as a continuation of the superheat table at very high pressures. It also serves as a continuation of a hypothetical compressed liquid table at these pressures. No actual compressed liquid table is included for subcritical pressures, as the data can be determined with sufficient accuracy from the saturation table. In addition to properties that have already been introduced, the steam tables include the property *specific entropy*, which has the symbol s and the units J/kg K. This is defined and explained in Chapter 14. The procedures for looking up its value in the tables are the same as for the properties u, h and v.

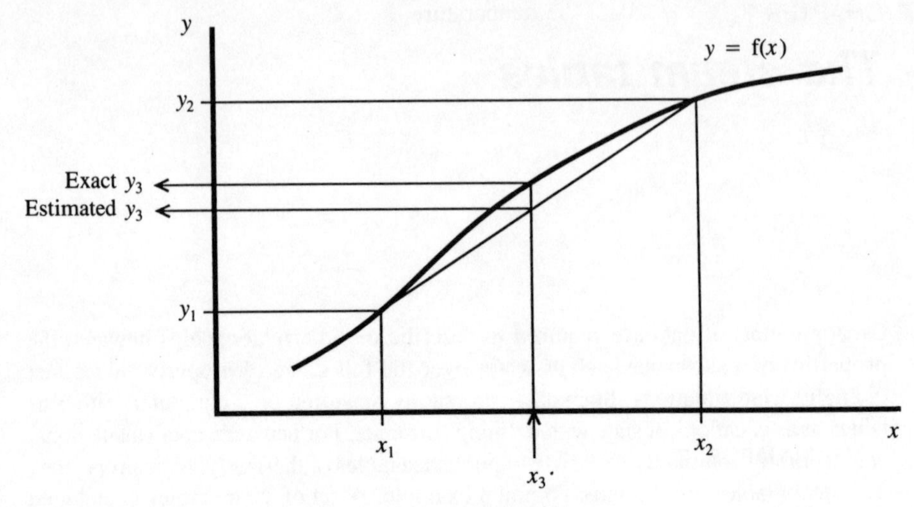

Figure 7.1 *Linear interpolation.*

7.2 Interpolation

Linear interpolation is used to find properties where the exact values of the known properties are not listed in the tables. The principle of the technique is illustrated in Figure 7.1. It can be seen by inspection of this figure that the estimated value of y_3, corresponding to x_3 when the coordinates (x_1, y_1) and (x_2, y_2) are known, is given by Equation (7.1). It is not advisable to try to memorize this equation, but rather to understand it and then apply it in an intuitive way. Linear interpolation from the tables becomes routine with a little practice.

$$y_3 = y_1 + \frac{x_3 - x_1}{x_2 - x_1}(y_2 - y_1). \tag{7.1}$$

7.3 Subcooled liquid (compressed liquid)

Liquid in the region to the left of the saturated liquid line in Figure 6.5 of Chapter 6 can be described as being subcooled with respect to the corresponding saturation temperature; or as being compressed with respect to the corresponding saturation pressure.

For most liquids at moderate pressures, the properties u and v depend mainly on temperature and vary little with pressure. Therefore, the corresponding properties of saturated liquid at the same temperature can usually be used with acceptable accuracy.[10] By definition, specific enthalpy depends on pressure. However, the specific enthalpy of subcooled or compressed liquid can be estimated readily from the saturation value at the same temperature, as follows:

$$h = u + pv$$

$$\approx u_f + pv_f = h_f - p_s v_f + pv_f.$$

Therefore,

$$h \approx h_f + v_f(p - p_s) \quad (7.2)$$

where h_f, v_f and p_s are determined at the *temperature* of the liquid and p is its pressure. The saturated liquid properties from Table A1 can thus be used to find the properties of subcooled liquid water.

EXAMPLE 7.1

Determine the specific enthalpy and the specific volume of water substance at 0.1 MPa and 10 °C.

SOLUTION

From the saturation table at a pressure of 0.1 MPa

$$t_s = 99.6 \text{ °C}.$$

Therefore, the water substance is subcooled. Alternatively, from the saturation table at a temperature of 10 °C

$$p_s = 0.001\ 23 \text{ MPa}$$

and the water is compressed.

From the saturation table at 10 °C

$$h_f = 42.0 \text{ kJ/kg}$$

$$v_f = 0.001\ 000 \text{ m}^3\text{/kg}.$$

As

$$h \approx h_f + v_f(p - p_s),$$

10. For water substance at the critical pressure, 22.12 MPa, and 50 °C, the error in specific internal energy would be about +1.6% and the error in specific volume about +0.9% if the saturation values at 50 °C (with a corresponding saturation pressure of 0.0123 MPa) were used. For pressures below the critical pressure, the errors would be correspondingly smaller.

$$h \approx 42.0 \times 10^3 \text{ [J kg}^{-1}] + 0.001\,000 \text{ [m}^3 \text{ kg}^{-1}] (0.1 - 0.001\,23) \times 10^6 \text{ [N m}^{-2}]$$

$$= 42\,099 \text{ J/kg}$$

Answer

$$= 42.1 \text{ kJ/kg.}$$

As

$$v \approx v_f,$$

Answer

$$v \approx 0.001\,000 \text{ m}^3\text{/kg.}$$

7.4 Saturated vapour

Table A1 of Appendix A is the saturation table for water and steam. The main thermodynamic properties are tabulated against saturation pressure and saturation temperature from the triple line to the critical point.

The saturation table is useful for finding the saturation temperature corresponding to a given pressure or the saturation pressure corresponding to a given temperature. It can also be used to find the specific internal energy, the specific enthalpy, or the specific volume for saturated liquid or dry saturated vapour, given the pressure or temperature. For convenience in evaluating the properties of mixtures of liquid and vapour, the property difference values u_{fg} and h_{fg} are also included.

EXAMPLE 7.2

Determine the specific internal energy and the specific volume of saturated liquid water and of dry saturated steam at 307 °C.

SOLUTION

From the saturation table for water,

at 300 °C

$$u_f = 1333.0 \text{ kJ/kg}$$

$$u_g = 2565.0 \text{ kJ/kg}$$

$$v_f = 0.001\,404 \text{ m}^3\text{/kg}$$

$$v_g = 0.021\,65 \text{ m}^3\text{/kg}$$

Note: Property values at a saturation temperature of 303.3 °C are also available in the saturation table. These could be used instead of those at 300 °C for the purposes of interpolation. This is a matter of numerical convenience. For the highest accuracy it would probably be better to use the values at 303.3 °C as this temperature is closer to 307 °C.

at 310 °C

$$u_f = 1388.1 \text{ kJ/kg}$$

$$u_g = 2549.1 \text{ kJ/kg}$$

$$v_f = 0.001\,448 \text{ m}^3/\text{kg}$$

$$v_g = 0.018\,33 \text{ m}^3/\text{kg}.$$

Therefore, at 307 °C

$$u_f = 1333.0 + \frac{307 - 300}{310 - 300}(1388.1 - 1333.0) \text{ kJ/kg}$$

$$= 1333.0 + 0.700\,(55.1) \text{ kJ/kg}$$

Answer

$$= 1371.6 \text{ kJ/kg}.$$

When interpolating to find the values of several properties, the same interpolation factor, 0.700 in the above expression for u_f, may be used several times. Therefore, it is worth while to evaluate the factor separately the first time, as above.

Similarly, at 307 °C

$$u_g = 2565.0 + 0.700\,(2549.1 - 2565.0) \text{ kJ/kg}$$

Answer

$$= 2553.9 \text{ kJ/kg}.$$

Also

$$v_f = 0.001\,404 + 0.700\,(0.001\,448 - 0.001\,404) \text{ m}^3/\text{kg}$$

Answer

$$= 0.001\,435 \text{ m}^3/\text{kg}$$

and

$$v_g = 0.021\,65 + 0.700\,(0.018\,33 - 0.021\,65)\ \text{m}^3/\text{kg}$$

Answer

$$= 0.019\,33\ \text{m}^3/\text{kg}.$$

EXAMPLE 7.3

If the temperature of saturated water and steam is 20 °C while its dryness is 75%, find its pressure, specific enthalpy and specific volume.

SOLUTION

From the tables

Answer

$$p_s = 0.002\,34\ \text{MPa}.$$

$$h = h_f + xh_{fg}$$

$$= 83.9 + 0.75(2454.3)\ \text{kJ/kg}$$

Answer

$$= 1924.6\ \text{kJ/kg}.$$

$$v = v_f + xv_{fg}$$

$$= 0.001\,002 + 0.75(57.837\,307)\ \text{m}^3/\text{kg}$$

Answer

$$= 43.38\ \text{m}^3/\text{kg}.$$

The same result would be obtained, to four significant digits, by ignoring v_f.

7.5 Superheated vapour

Table A2 of Appendix A is the superheat table. It includes data for superheated steam at pressures from the triple point pressure to the critical pressure, and at temperatures from the saturation value corresponding to each pressure up to a maximum of from 350 °C to 700 °C, depending on the pressure level. The data tabulated against p

and t are h, s and v. If values of u are required, they can be calculated from h, p and v. The first column of data in the superheat table consists of the saturation temperature and the properties h, s and v of dry saturated vapour at each pressure.

Two main situations arise where interpolation is required. Example 7.4 illustrates the first, where p and t are both given. The second is where either p or t is known together with one other property, which could be h, u, v or s. Example 7.5 is of this type. Double interpolation is often necessary, as in Example 7.4.

EXAMPLE 7.4

Determine the specific enthalpy of steam at 0.77 MPa and 380 °C.

SOLUTION

From the superheat table at 0.7 MPa

at 350 °C $\qquad h = 3164.3$ kJ/kg

at 400 °C $\qquad h = 3269.0$ kJ/kg

Therefore, at 0.7 MPa and 380 °C

$$h = 3164.3 + \frac{380 - 350}{400 - 350}(3269.0 - 3164.3) \text{ kJ/kg}$$

$$= 3227.1 \text{ kJ/kg.}$$

From the superheat table at 0.792 MPa

at 350 °C $\qquad h = 3162.5$ kJ/kg

at 400 °C $\qquad h = 3267.6$ kJ/kg.

Therefore, at 0.792 MPa and 380 °C

$$h = 3162.5 + \frac{380 - 350}{400 - 350}(3267.6 - 3162.5) \text{ kJ/kg}$$

$$= 3225.6 \text{ kJ/kg.}$$

Hence, at 0.77 MPa and 380 °C

$$h = 3227.1 + \frac{0.77 - 0.7}{0.792 - 0.7}(3225.6 - 3227.1) \text{ kJ/kg}$$

Answer

$$= 3226.0 \text{ kJ/kg.}$$

EXAMPLE 7.5

Determine the temperature and specific enthalpy of steam that has a pressure of 1.5 MPa and a specific volume of 0.1900 m^3/kg.

SOLUTION

From the information given, the steam could be either wet or superheated.

From the saturation table at 1.5 MPa

$$v_g = 0.1317\ m^3/kg.$$

As $v > v_g$, the steam is superheated.

From the superheat table at 1.5 MPa

at 350 °C $\qquad v = 0.1865\ m^3/kg$

at 400 °C $\qquad v = 0.2029\ m^3/kg.$

Therefore, at 1.5 MPa and $v = 0.1900\ m^3/kg$

$$t = 350 + \frac{0.1900 - 0.1865}{0.2029 - 0.1865}(400 - 350)\ °C$$

$$t = 350 + 0.2134(50)\ °C$$

Answer

$$= 360.7\ °C.$$

The specific enthalpy value can now be found at 1.5 MPa and $v = 0.1900\ m^3/kg$ ($t = 360.7$ °C).

Using values from the superheat table at 1.5 MPa

at 0.1900 m^3/kg $\qquad h = 3148.7 + 0.2134\ (3256.6 - 3148.7)\ kJ/kg$

Answer

$$= 3171.7\ kJ/kg.$$

7.6 Substance at supercritical pressure

Data for water substance at supercritical pressures are included in Table A3. The pressure range is from the critical value of 22.12 MPa to 100 MPa. The temperature range is from 0 °C to 800 °C. The procedure for looking up this table is the same as for Table A2.

EXAMPLE 7.6

Determine the specific volume and the specific enthalpy of
(a) saturated liquid water at 50 °C
(b) compressed liquid water at 50 °C and 5 MPa
(c) water substance at 50 °C and 50 MPa
(d) water substance at 500 °C and 50 MPa.

SOLUTION

(a) From the saturation table at 50 °C

Answer (a)

$$v_f = 0.001\,012 \text{ m}^3/\text{kg}$$

Answer (a)

$$h_f = 209.3 \text{ kJ/kg}.$$

(b) At 50 °C and 5 MPa

$$v \approx v_f \text{ at } 50\text{ °C}.$$

Therefore

Answer (b)

$$v \approx 0.001\,012 \text{ m}^3/\text{kg}.$$

From the saturation table at 50 °C

$$p_s = 0.0123 \text{ MPa}$$

$$h \approx h_f + (p - p_s)v_f$$

$$= 209.3\ [\text{kJ kg}^{-1}] + (5 - 0.0123)10^3\ [\text{kPa}]\ (0.001\,012)\ [\text{m}^3\ \text{kg}^{-1}].$$

Therefore

Answer (b)

$$h \approx 214.3 \text{ kJ/kg}.$$

(c) From the table for water substance at supercritical pressures at 50 °C and 50 MPa

Answer (c)

$$h = 251.9 \text{ kJ/kg}$$

Answer (c)

$$v = 0.000\,991 \text{ m}^3/\text{kg}.$$

(d) From the table for water substance at supercritical pressures at 500 °C and 50 MPa

Answer (d)

$$h = 2723.0 \text{ kJ/kg}$$

Answer (d)

$$v_f = 0.003\,882 \text{ m}^3/\text{kg}.$$

The four state points corresponding to (a)–(d) are sketched on a t–h diagram in Figure 7.2.

7.7 Practical tips

- Make sure you can sketch the saturated liquid and dry saturated vapour lines on a p–v diagram and on a T–u or T–h diagram. In addition, you should be able to sketch constant temperature lines on the p–v diagram and constant pressure lines on the T–u and T–h diagrams. Keep these diagrams in mind as you use the steam tables.
- For saturated steam you need to be able to use the expression $h = h_f + xh_{fg}$. Exactly the same form of expression applies for u and v (and s). Also, if h, h_f and h_{fg} or h_g are known, the dryness fraction can be found by rearranging the expression.
- In the compressed or subcooled liquid region, the properties u and v depend on the temperature and are not greatly influenced by the pressure. It is usually accurate enough to use the values for saturated liquid at the same *temperature*. Specific enthalpy values, however, may need to be corrected for the excess pressure above the saturation value.
- You need to develop the skill of performing linear interpolation. This is quickly developed by doing some examples. This should not be a matter of memorizing the formula, but of understanding the method.

7.8 Summary

The steam tables have been introduced and explained with some worked examples. The student should be able to

- *apply*
  — the linear interpolation technique to data from the steam tables

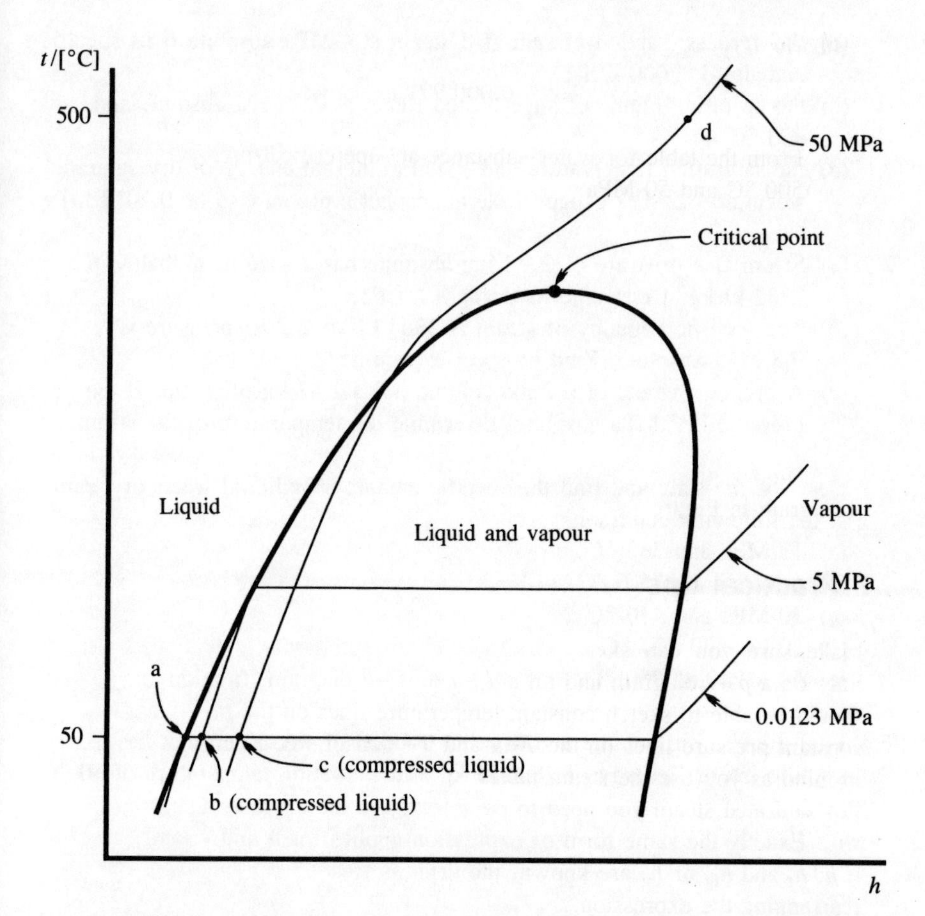

Figure 7.2 *Pressure versus specific enthalpy diagram showing the state points referred to in Example 7.6. The distances between points a and b and between points a and c are exaggerated.*

- *use*
  — the steam tables to determine the thermodynamic properties p, T, u, h and v for saturated liquid, dry saturated vapour, liquid and vapour mixtures, subcooled or compressed liquid, superheated vapour and substance at supercritical pressures.

7.9 Self-assessment questions

7.1 Use the steam tables to find the following properties of water and steam:
  (a) the specific enthalpy and the specific volume of saturated liquid water at 40 °C

(b) the dryness fraction of saturated steam at 4 MPa absolute if its specific enthalpy is 2600 kJ/kg
(c) the specific volume of superheated steam at 0.5 MPa absolute and 330 °C
(d) the saturation temperature and specific internal energy of dry saturated steam at 4.2 MPa gauge. Take atmospheric pressure to be 0.1013 MPa.

7.2 (a) Steam at a pressure of 2.9 MPa absolute has a specific enthalpy of 2732 kJ/kg. Determine its dryness fraction.
(b) The specific enthalpy of steam is 3364 kJ/kg and its pressure is 0.8 MPa absolute. Find its specific volume.
(c) A pressure vessel of 0.2 m^3 volume holds 1.43 kg of steam. If the pressure is 2 MPa absolute, determine the temperature of the steam.

7.3 Describe the state and find the specific enthalpy for liquid water or steam at the following conditions:
(a) 13 MPa and 360 °C
(b) 6 MPa and 275.6 °C
(c) 30 MPa and 410 °C
(d) 20 bar and 100 °C
(e) 50 MPa and 250 °C.

7.4 Find the specific volume and dryness fraction of wet steam that has a specific internal energy of 2000 kJ/kg at a saturation temperature of 50 °C.

CHAPTER 8

Ideal gases

The theory of ideal gases is presented in this chapter. This includes the ideal gas equation, relationships for the internal energy and enthalpy of ideal gases, some relationships for processes undergone by ideal gases and the relationships between the parameters that describe particular ideal gases. Values for these parameters are included in Appendix B for some common gases.

The practical advantage of taking gases to be ideal is that a simple equation of state with only one constant applies and very few data are required to describe a particular gas over a wide range of its possible thermodynamic equilibrium states. If it is also assumed that the specific heats c_p and c_v are constant, then changes in the specific internal energy or the specific enthalpy can be calculated simply without the need for thermodynamic tables.

A *permanent gas* is a gas that has a critical temperature below ambient temperature. Most of the permanent gases, such as air or hydrogen, can be regarded as 'ideal gases' in many engineering situations, especially at temperatures well above their respective critical temperatures and pressures well below their respective saturation pressures at given temperatures. At very low pressures all gases or vapours, including non-permanent gases such as water vapour, approach the ideal gas model.

8.1 The ideal gas equation

The ideal gas equation was originally developed from experimental observations of real gases at low pressures. For a closed system containing a gas it was found that, at a given temperature, the pressure was very nearly inversely proportional to the specific volume. Thus, at sufficiently low pressures isothermal lines formed hyperbolas on a $p-v$ diagram.

Also, a very satisfactory linear temperature scale could be constructed based on the pressure of a constant volume system that contained a fixed mass of gas. It was found that, provided the pressure was sufficiently low, such temperature scales based on different gases and on the freezing and boiling points of water at standard atmospheric pressure as fixed reference temperatures were in excellent agreement with one another. Furthermore, *constant volume gas thermometer* scales based on different gases predicted the same temperature (about -273.15 °C) at which the pressure of the constant volume of gas would become zero. A constant volume gas thermometer temperature scale agrees with the modern absolute temperature scale (which is defined in Chapter 13, section 13.7) as the pressure in the constant volume gas thermometer (at the highest measurement temperature) tends to zero. Therefore, the pressure of a constant volume of a real gas at low pressure is approximately proportional to its absolute temperature.

It has thus been established experimentally that the behaviour of real gases at low pressure can be described accurately by an equation, in which pressure is inversely proportional to specific volume at constant temperature and proportional to the absolute temperature at constant specific volume. This equation has the form

$$pv = RT \tag{8.1}$$

and

$$pV = mRT \tag{8.2}$$

where R is a constant for the gas known as the *specific gas constant*. The specific gas constant can be evaluated experimentally for any given gas at low pressure by measuring p, V, T and m. Equations (8.1) and (8.2) are two forms of the equation known as the *ideal gas equation of state*. An *ideal gas* is a gas that obeys the ideal gas equation.

Another important experimental observation that had historical importance is that at a given volume, temperature and pressure the number of kilogram moles is very nearly the same no matter what gas is present, provided that the pressure is low. As the pressure tends to zero, the number of kilogram moles will tend to be exactly the same no matter what gas is present:

$$n = \frac{m}{\bar{m}} = \text{const} \quad \text{for all gases at fixed } p,\ V \text{ and } T \tag{8.3}$$

where n is the number of kilogram moles present and $\bar{m}$ is the molar mass, which has the units kg/kmol. Hence

$$n = \frac{pV/RT}{\bar{m}} = \frac{pV}{\bar{m}RT} = \text{const} \quad \text{for all gases at fixed } p,\ V \text{ and } T. \tag{8.4}$$

Hence

$$\bar{m}R = \text{const} \quad \text{for all gases.} \tag{8.5}$$

Therefore, the specific gas constant multiplied by the molar mass, $\bar{m}$, has the same value for all gases. This is known as the *universal gas constant*. Hence

$$\bar{R} = \bar{m}R \tag{8.6}$$

where $\bar{R}$ is the universal gas constant.

The universal gas constant

$$\bar{R} = 8.3144 \text{ kJ/kmol K.}$$

The ideal gas equation can also be written in the following forms:

$$pV = \frac{m}{\bar{m}}\bar{R}T \tag{8.7}$$

or

$$pV = n\bar{R}T. \tag{8.8}$$

EXAMPLE 8.1

Case study: air pressure in a sealed room

An empty cold room has a volume of 23 m^3. If the pressure inside is 1 bar and the temperature is -10 °C, what mass of air does it contain? In order to minimize the need for refrigeration, it would be desirable to minimize any loss of cold air and any gain of warm air, but sealing the cold room completely could cause other problems. If the cold room were sealed and the temperature within it were allowed to rise to 20 °C, what would be the pressure inside? Calculate the net force that would act on a door of area 2.5 m^2 when the temperature in the cold room had reached 20 °C and the pressure outside remained constant at 1 bar. Comment on the problems due to the cold room being sealed in these circumstances. (For air, $R = 0.2870$ kJ/kg K.)

SOLUTION

$$pV = mRT.$$

Therefore

$$m = \frac{pV}{RT}$$

$$= \frac{10^5 \times 23 \ [\text{N m}^{-2}\text{m}^3]}{287.0 \times 263.15 \ [\text{J kg}^{-1}\text{K}^{-1}\text{K}]}$$

Answer

$$= 30.45 \text{ kg.}$$

Since m and R are constants

$$\frac{p_1 V_1}{T_1} = \frac{p_2 V_2}{T_2}$$

and since $V_1 = V_2$

$$p_2 = p_1 \frac{T_2}{T_1}$$

Answer

$$= 1\ [\text{bar}]\ \frac{293.15\ [\text{K}]}{263.15\ [\text{K}]} = 1.114\ \text{bar}.$$

The net force that acts on the door, as shown in Figure 8.1, is given by

$$F_{\text{net}} = (p_{\text{i}} - p_{\text{o}})A$$

$$= (1.114 - 1)\ 10^5\ [\text{N m}^{-2}]\ 2.5\ [\text{m}^2]$$

Answer

$$= 28\ 500\ \text{N} = 28.50\ \text{kN}.$$

One problem due to the room being sealed in the circumstances mentioned would be the high force that would act on the door. Another problem, which could have safety implications for personnel, would be the high force necessary to open the door from the inside if it were hinged to open inwards. A small vent would eliminate these problems. The small

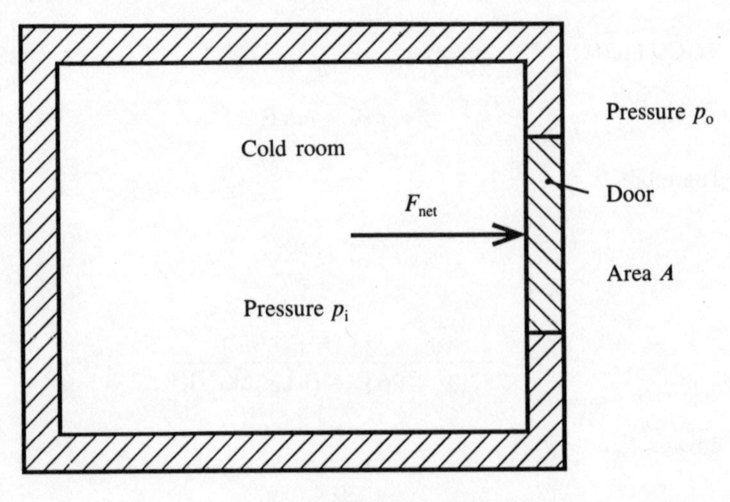

Figure 8.1

additional refrigeration load associated with this would probably be considered well justified.

8.2 Joule's law

Joule's law states that the specific internal energy of an ideal gas depends only on its temperature:

$$u = \mathrm{f}(T). \tag{8.9}$$

Joule's law was originally based on an experiment carried out by James Prescott Joule. Strictly, it only applies for real gases as the pressure tends to zero. In general, the specific internal energy will also depend (usually weakly) on a second intensive property such as pressure. Therefore, Joule's law is strictly true (and thus a 'law') only for ideal gases. Joule's law has a number of consequences that simplify calculations involving ideal gases, as follows:

- From the definition of specific enthalpy

$$h = u + pv = u + RT = \mathrm{f}'(T). \tag{8.10}$$

  That is, the specific enthalpy of an ideal gas also depends only on its temperature.
- For an ideal gas the definitions of the specific heat capacities (Equations (6.12) and (6.13), Chapter 6) become

$$c_v = \frac{\mathrm{d}u}{\mathrm{d}T} \tag{8.11}$$

$$c_p = \frac{\mathrm{d}h}{\mathrm{d}T}. \tag{8.12}$$

  In the case of the definition of the specific heat at constant volume (Equation (8.11)), there is no requirement that the derivative of the specific internal energy with respect to temperature is evaluated at constant specific volume. This is because, from Joule's law, the internal energy only depends on the temperature, whether or not the specific volume is constant. In the case of the specific heat at constant pressure (Equation (8.12)), there is no requirement that the derivative of the specific enthalpy with respect to temperature is evaluated at constant pressure. This is because the specific enthalpy only depends on the temperature irrespective of whether the pressure is constant or not.

8.3 Internal energy and enthalpy differences

The specific heat values themselves depend, in general, on the temperature, even for ideal gases. However, it is often the case that, over the temperature range of interest, the changes in the specific heat values can be neglected. Even where the changes in specific heat values cannot be neglected, it is often sufficiently accurate to use the mean specific heat values for the temperature range.

From Equation (8.11),

$$du = c_v \, dT. \tag{8.13}$$

If c_v is assumed to be constant, then, for any process between equilibrium states 1 and 2, Equation (8.13) can be integrated readily to give

$$\Delta u_{1-2} = c_v(T_2 - T_1). \tag{8.14}$$

Similarly

$$dh = c_p \, dT \tag{8.15}$$

and, if c_p is assumed to be constant, for any process between equilibrium states 1 and 2

$$\Delta h_{1-2} = c_p(T_2 - T_1). \tag{8.16}$$

EXAMPLE 8.2

A system contains 2.3 kg of hydrogen at 1.05 bar absolute and 21 °C. After a process, A, it reaches an equilibrium pressure of 2.7 bar absolute at the original temperature. After a further process, B, it reaches an equilibrium state at 2.7 bar absolute and a temperature of 37 °C. If the specific internal energy at the initial state is 213.1 kJ/kg, determine

(a) the internal energy and enthalpy of the system at the initial state
(b) the internal energy and enthalpy of the system after process A
(c) the internal energy and enthalpy after process B.

Hydrogen can be regarded as an ideal gas – use the ideal gas data from Appendix B.

SOLUTION

From Appendix B, c_p = 14.27 kJ/kg K, c_v = 10.15 kJ/kg K and R = 4.1244 kJ/kg K.

$$U_1 = mu_1$$

$$= 2.3 \text{ [kg] } 213.1 \text{ [kJ kg}^{-1}\text{]}$$

Answer (a)

$$= 490.13 \text{ kJ}.$$

$$H_1 = U_1 + p_1 V_1$$

$$= U_1 + mRT_1$$

$$= 490.13 \text{ [kJ]} + 2.3 \text{ [kg]} \; 4.1244 \text{ [kJ kg}^{-1}\text{K}^{-1}] \; (21 + 273.15) \text{ [K]}$$

Answer (a)

$$= 3280.5 \text{ kJ}.$$

After process A (Figure 8.2),

$$T_2 = T_1.$$

Therefore,

Answer (b)

$$U_2 = U_1 = 490.13 \text{ kJ}$$

and

Answer (b)

$$H_2 = H_1 = 3280.5 \text{ kJ}.$$

$$U_3 = U_1 + mc_v(T_3 - T_1)$$

$$= 490.13 \text{ [kJ]} + 2.3 \text{ [kg]} \; 10.15 \text{ [kJ kg}^{-1}\text{K}^{-1}] \; (37 - 21) \text{ [K]}$$

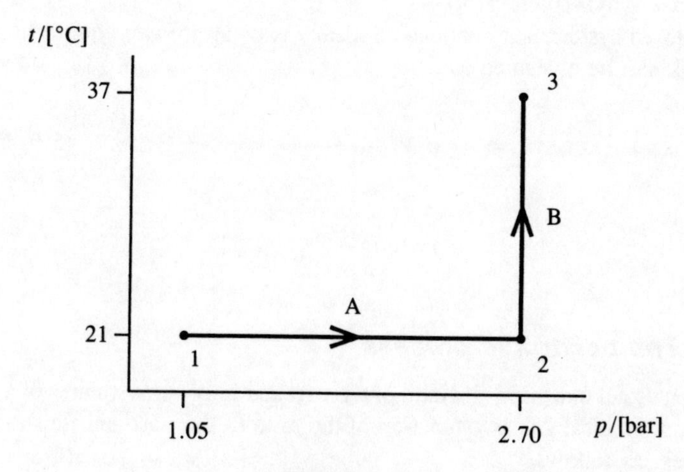

Figure 8.2

Answer (c)

$$= 863.65 \text{ kJ}.$$

$$H_3 = H_2 + mc_p(T_3 - T_2)$$

$$= 3280.5 \text{ [kJ]} + 2.3 \text{ [kg] } 14.27 \text{ [kJ kg}^{-1}\text{K}^{-1}\text{] } (37 - 21) \text{ [K]}$$

Answer (c)

$$= 3805.6 \text{ kJ}.$$

8.4 Processes of an ideal gas

There are several types of processes of ideal gases that are frequently encountered in engineering situations. These are described in the following sections.

8.4.1 The isothermal process

In an isothermal process the temperature is constant. Therefore, from the ideal gas equation, Equation (8.1), the following relationship applies:

$$pv = \text{const} \qquad (8.17)$$

or, if the mass is constant, as in the case of a closed system, Equation (8.2) yields

$$pV = \text{const}. \qquad (8.18)$$

Each of these equations describes a hyperbola. An isothermal process of an ideal gas is thus a hyperbolic process.

If a closed system that contains an ideal gas undergoes an equilibrium process, the work can be evaluated as

$$W = p_1 V_1 \ln \frac{p_2}{p_1} \qquad \text{(5.5, Chapter 5)}$$

$$= mRT \ln \frac{p_2}{p_1}. \qquad (8.19)$$

8.4.2 The polytropic process

It is often useful to express the ratio of the start and finish temperatures of a polytropic process of an ideal gas as a function of the ratio of the start and finish volumes or pressures, as follows.

From the polytropic relation

$$p_1 V_1^n = p_2 V_2^n.$$

Hence

$$\frac{p_2}{p_1} = \left(\frac{V_1}{V_2}\right)^n \tag{8.20}$$

and

$$\frac{V_1}{V_2} = \left(\frac{p_2}{p_1}\right)^{1/n}. \tag{8.21}$$

From the ideal gas equation and for a closed system

$$\frac{p_1 V_1}{T_1} = \frac{p_2 V_2}{T_2}. \tag{8.22}$$

Hence

$$\frac{T_2}{T_1} = \frac{p_2 V_2}{p_1 V_1}. \tag{8.23}$$

Substituting for the pressure ratio,

$$\frac{T_2}{T_1} = \left(\frac{V_1}{V_2}\right)^{n-1}. \tag{8.24}$$

Substituting for the volume ratio,

$$\frac{T_2}{T_1} = \left(\frac{p_2}{p_1}\right)^{(n-1)/n}. \tag{8.25}$$

If a closed system that contains an ideal gas undergoes a polytropic process, the work can be evaluated as

$$W = \frac{p_2 V_2 - p_1 V_1}{n - 1} \qquad \text{(5.7, Chapter 5)}$$

$$= \frac{mR(T_2 - T_1)}{n - 1}. \tag{8.26}$$

8.4.3 *The adiabatic equilibrium process*

It is found experimentally that a closed system containing a gas at low pressure obeys the polytropic rule with a characteristic exponent when it undergoes an adiabatic equilibrium process. The exponent is known as the *adiabatic index*, γ:

$$pV^{\gamma} = \text{const.} \tag{8.27}$$

For an adiabatic equilibrium process of an ideal gas, the work is given by Equation (5.7), Chapter 5, where $n = \gamma$:

$$W = \frac{p_2 V_2 - p_1 V_1}{\gamma - 1} \tag{8.28}$$

$$= \frac{mR(T_2 - T_1)}{\gamma - 1}. \tag{8.29}$$

8.5 Relationships between the ideal gas parameters

It will be shown in Chapter 10, section 10.5, that for an ideal gas the adiabatic index equals the ratio of the specific heat values:

$$\gamma = \frac{c_p}{c_v}. \tag{8.30}$$

For any two states of an ideal gas that are close enough together that any change in c_p can be neglected

$$h_2 - h_1 = c_p(T_2 - T_1) \tag{8.31}$$

$$(u_2 - u_1) + (p_2 v_2 - p_1 v_1) = c_p(T_2 - T_1) \tag{8.32}$$

$$c_v(T_2 - T_1) + R(T_2 - T_1) = c_p(T_2 - T_1). \tag{8.33}$$

Therefore

$$R = c_p - c_v. \tag{8.34}$$

Hence, only two of the four parameters, c_p, c_v, R and γ are independent.

Ideal gas parameters for air

For approximate calculations with air, the following ideal gas parameters may be assumed:

R = 0.2870 kJ/kg K $\qquad \gamma$ = 1.4

c_p = 1.005 kJ/kg K $\qquad c_v$ = 0.718 kJ/kg K.

EXAMPLE 8.3

Case study: foot pump

Within one stroke of a foot pump that is being used to inflate a tyre, air at 1 bar and 15 °C is compressed adiabatically in a quasi-equilibrium process to 4 bar. While this occurs, the cylinder is closed because the pump's intake valve had shut when the compression process began and the valve of the tyre has not yet opened to allow the compressed air out of the pump. Determine the temperature, the compression ratio (i.e. V_1/V_2) and the work done on the air per unit mass at the moment when the pressure of 4 bar is reached.

SOLUTION

The process is illustrated in Figure 8.3. For air, γ = 1.4 and R = 0.2870 kJ/kg K.

$$\frac{T_2}{T_1} = \left(\frac{p_2}{p_1}\right)^{(\gamma-1)/\gamma}$$

$$T_2 = (273.15 + 15)\left(\frac{4}{1}\right)^{0.4/1.4} \text{ [K]}$$

$$= 428.19 \text{ K}$$

Answer

$$t_2 = T_2 - 273.15 \text{ [K]} = 155.0 \text{ °C}.$$

$$\frac{p_1 V_1}{T_1} = \frac{p_2 V_2}{T_2}$$

Answer

$$\frac{V_1}{V_2} = \frac{p_2}{p_1}\frac{T_1}{T_2} = \frac{4}{1}\left(\frac{288.15}{428.19}\right) = 2.692.$$

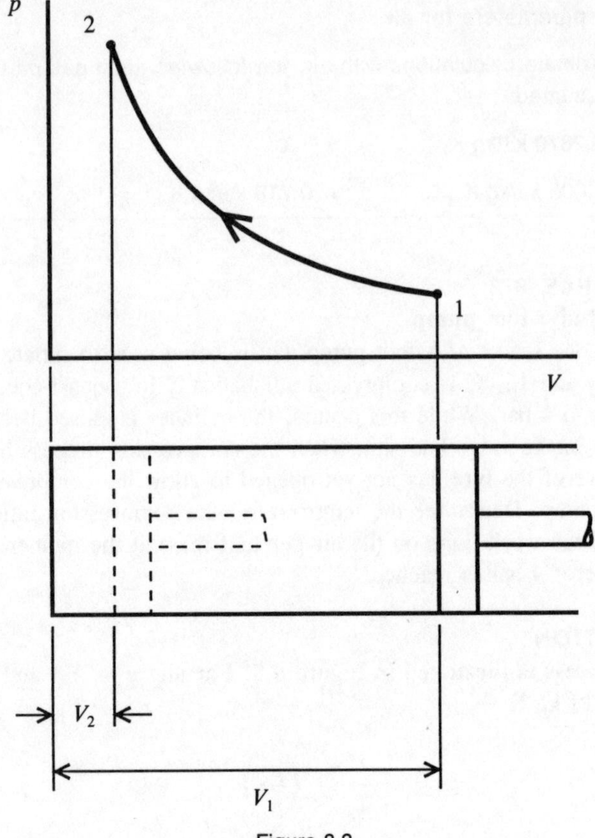

Figure 8.3

The work done on the air per unit mass is

$$w = \frac{R(T_2 - T_1)}{\gamma - 1}$$

$$= \frac{0.2870(155 - 15)}{0.4} \text{ [kJ kg}^{-1}\text{K}^{-1}\text{] [K]}$$

Answer

$$= 100.5 \text{ kJ/kg}.$$

8.6 Practical tips

- In the ideal gas equation and any relationships based on it, the temperature must be the absolute temperature.

- The pressure in the ideal gas equation and any relationships based on it must be the absolute pressure.
- Do not try to memorize the expressions for the ratio of the start and finish temperatures of a polytropic process (Equations (8.24) and (8.25). Instead, make sure that you can derive them quickly, when required, from the ideal gas relationship pV/T = const and the polytropic relationship pV^n = const.

8.7 Summary

The theory of ideal gases has been introduced in this chapter. The ideal gas equation has been presented and explained. The specific gas constant, the universal gas constant and the adiabatic index have been introduced. Joule's law has been stated and expressions have been given for calculating changes in the internal energy or enthalpy of an ideal gas. Property relations and work expressions for isothermal, polytropic and adiabatic processes of an ideal gas have been derived. The relationships between the parameters that define an ideal gas have been described.

The student should be able to

- *define*
  — a permanent gas, an ideal gas
- *explain*
  — a constant volume gas thermometer, the specific gas constant, the universal gas constant, the adiabatic index, the relationships between the ideal gas parameters
- *write*
  — the ideal gas equation in forms that include the specific gas constant or the universal gas constant, defining expressions for the specific heats at constant pressure and at constant volume for an ideal gas
- *state*
  — Joule's law
- *show*
  — that, given Joule's law, the specific enthalpy of an ideal gas depends only on its temperature
- *derive*
  — expressions for the ratio of the start and finish temperatures of a polytropic process of an ideal gas in terms of the start and finish volumes or pressures, the relationship between the specific gas constant and the specific heats of an ideal gas
- *perform calculations using*
  — the ideal gas equation and ideal gas relationships; the specific heats of

ideal gases; work expressions for ideal gases for isothermal, polytropic or adiabatic processes.

8.8 Self-assessment questions

8.1 A quantity of ideal gas has an initial gauge pressure of 0.15 MPa and a volume of 0.2 m^3. If it is compressed to a gauge pressure of 0.7 MPa while the temperature is maintained constant, determine the final volume. Take atmospheric pressure as 0.1 MPa.

8.2 A mass of 5 kg of an ideal gas has an initial volume of 0.07 m^3 and an initial temperature of 16 °C. If the temperature is raised to 280 °C while the pressure is maintained constant, determine the final volume of the gas. What is the change in enthalpy induced by this process? Take c_p as 1.065 kJ/kg K.

8.3 A mass of ideal gas has an initial temperature of 150 °C, absolute pressure of 14 bar and volume of 0.015 m^3. It has a specific gas constant of 0.295 kJ/kg K and a specific heat at constant pressure of 1.01 kJ/kg K. If it is expanded until its final pressure and volume are 3 bar absolute and 0.06 m^3 respectively, determine:
(a) the mass of gas present
(b) the final temperature

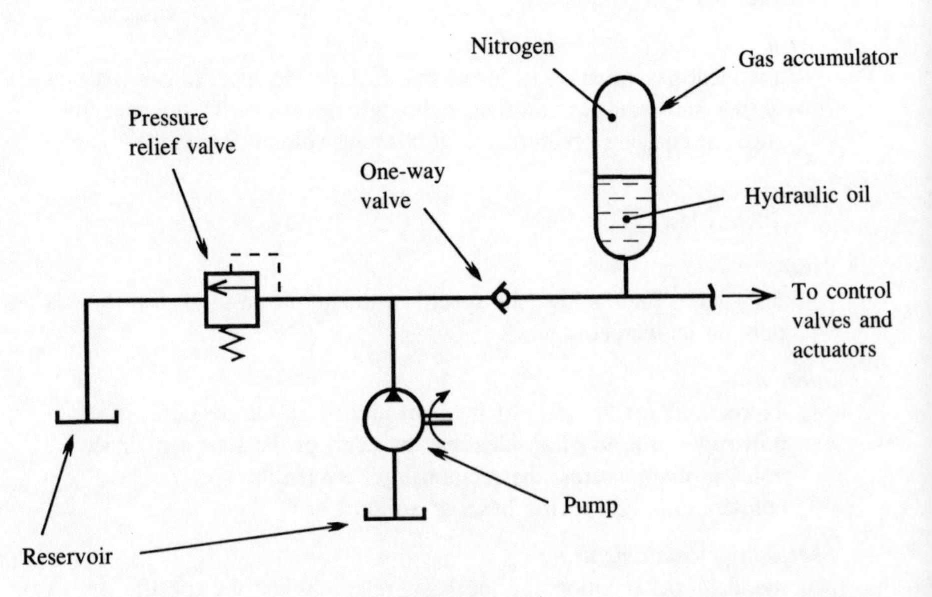

Figure 8.4

(c) the specific volumes before and after expansion
(d) the change in the internal energy.

8.4 Figure 8.4 is a typical schematic representation of part of a hydraulic circuit. It could, for instance, be part of a machine for crushing old cars. A gas accumulator in the hydraulic system is used to provide for short duration flow rate demands in excess of the pump output. If the gas in the accumulator initially occupies a volume of 0.355 m^3 and has a pressure of 0.96 MPa absolute, determine

(a) the work done on the gas when the pressure is increased to 8 MPa absolute due to hydraulic oil entering the accumulator and reducing the gas volume
(b) the work done by the gas when it subsequently expands to a pressure of 3.5 MPa absolute, pushing hydraulic oil back into the circuit.

Assume that the first process, which occurs slowly, is hyperbolic (for an ideal gas an isothermal process is a hyperbolic process). Assume that the second process, which occurs relatively quickly and is almost adiabatic, is polytropic with an exponent of 1.38.

CHAPTER 9

The mass balance equation

In this chapter the mass balance equation, which is the mathematical form of the principle of conservation of mass, is presented. It is shown how this can be applied to closed systems and to open systems. The mass balance equation also serves as preparation for the steady flow energy equation, which is the subject of Chapter 11.

9.1 Steady flows, states and systems

A *steady flow rate* is a flow rate that does not vary with time. A *non-steady (or unsteady) flow rate* is one that varies with time. The adjectives *steady* and *non*-steady (or *unsteady*) can be applied to other nouns in thermodynamics with the same sense: an item described as *steady* is invariant with time and an item described as *unsteady* varies with time. Thus, a *steady state* is a state that does not change with time. A *steady system* is a system of fixed size and shape within which the mechanical and thermodynamic properties are invariant with time at every point.

9.2 The mass balance equation

From the principle of conservation of mass (section 4.1, Chapter 4), a mass balance equation for a general system, as shown in Figure 9.1, can be expressed as follows:

$$\begin{pmatrix}\text{mass that enters}\\ \text{through the}\\ \text{boundary}\end{pmatrix} - \begin{pmatrix}\text{mass that leaves}\\ \text{through the}\\ \text{boundary}\end{pmatrix} = \begin{pmatrix}\text{increase in}\\ \text{mass of}\\ \text{the system}\end{pmatrix}. \tag{9.1}$$

Alternatively, a time rate equation can be written as follows:

$$\begin{pmatrix}\text{rate at which mass}\\ \text{enters through}\\ \text{the boundary}\end{pmatrix} - \begin{pmatrix}\text{rate at which mass}\\ \text{leaves through}\\ \text{the boundary}\end{pmatrix} = \begin{pmatrix}\text{rate of change}\\ \text{of mass within}\\ \text{the system}\end{pmatrix} \tag{9.2}$$

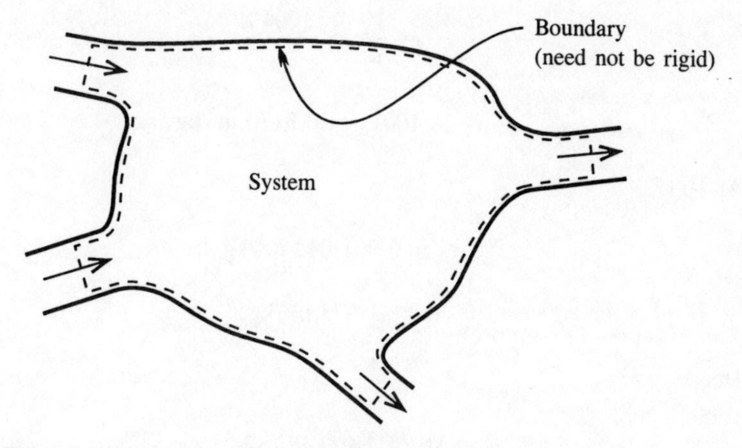

Figure 9.1 *A general system with non-steady flow rates inwards and outwards.*

$$\sum_{\text{in}} \dot{m} - \sum_{\text{out}} \dot{m} = \frac{\mathrm{d}m_{\text{sys}}}{\mathrm{d}\boldsymbol{t}} \tag{9.3}$$

where the symbol $\boldsymbol{t}$ represents time.

9.3 Closed systems

If a system is closed then

$$m_{\text{sys}} = \text{const} \tag{9.4}$$

and

$$\frac{\mathrm{d}m_{\text{sys}}}{\mathrm{d}\boldsymbol{t}} = 0. \tag{9.5}$$

EXAMPLE 9.1

Water at its critical temperature and pressure fills a closed system. It has a mass of 5 kg. If the volume increased by a factor of 100 and the temperature fell to 100 °C, what mass of liquid water would be present?

SOLUTION

From the table for saturated steam at t_c and p_c (where $v_{fg} = 0$)

$$v_c = 0.003\ 170\ \text{m}^3/\text{kg}.$$

Since mass is conserved and $m = V/v$

$$\frac{V_1}{v_c} = \frac{V_2}{v_2} = \frac{100V_1}{v_c}.$$

$$\therefore\ v_2 = 100v_c = 0.3170\ \text{m}^3/\text{kg}.$$

At 100 °C

$$v_f = 0.001\ 044\ \text{m}^3/\text{kg}$$

$$v_g = 1.673\ \text{m}^3/\text{kg}.$$

Hence

$$v_{fg} = 1.672\ \text{m}^3/\text{kg}.$$

$$x_2 = \frac{v_2 - v_f}{v_{fg}} = \frac{0.3170 - 0.001\ 044}{1.672} = 0.188\ 97$$

Answer

$$m_{liq} = m(1 - x_2) = 5(1 - 0.188\ 97)\ \text{kg} = 4.055\ \text{kg}.$$

9.4 Open systems

A uniform velocity is a velocity that has the same value at all points over a specified area. *A uniform flow* is a flow that has uniform velocity. For instance, the flow in a pipe is often assumed to be uniform even though, in reality, the velocity of the fluid reduces to zero at the pipe walls.

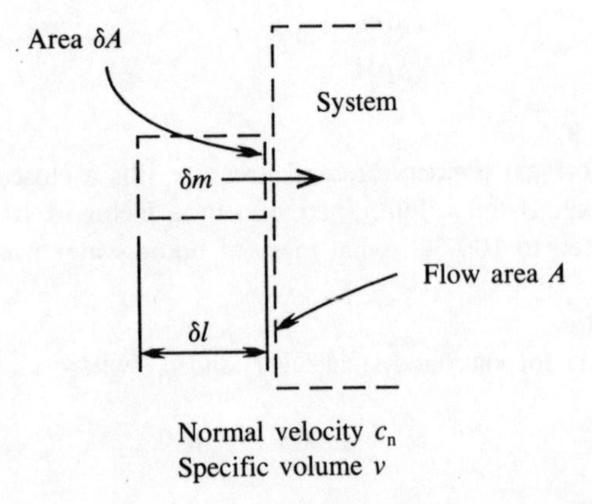

Figure 9.2 *A small element of mass about to enter a system.*

Figure 9.2 refers to a general situation where a substance crosses a system boundary. It shows a small element of mass δm that enters the system in time $\delta \boldsymbol{t}$ over an area δA with a velocity c_n normal to the boundary and a specific volume v. The mass flow rate over area δA is given by

$$\delta \dot{m} = \frac{\delta A \delta l}{v} \frac{1}{\delta \boldsymbol{t}}.$$

But

$$\frac{\delta l}{\delta \boldsymbol{t}} = c_n$$

and so

$$\delta \dot{m} = \frac{c_n}{v} \delta A. \tag{9.6}$$

Over a given flow area, A, of the boundary

$$\dot{m} = \int \frac{c_n}{v} \, dA. \tag{9.7}$$

If the specific volume, v, is constant over the area

$$\dot{m} = \frac{1}{v} \int c_n \, dA. \tag{9.8}$$

If, in addition, c_n is uniform over the area

$$\dot{m} = \frac{c_n A}{v}. \tag{9.9}$$

Equation (9.9) is important in fluid mechanics as well as in thermodynamics. In the area of fluid mechanics it is common to use the property density in preference to its inverse, specific volume. Equation (9.9) can thus be written in the form of Equation (9.10):

$$\dot{m} = \rho c_n A \tag{9.10}$$

where

$$\rho = \text{density} = \frac{1}{v}. \tag{9.11}$$

If the system is steady then the volume, the mass and all properties within the system are invariant with time. Hence

$$\frac{dm_{sys}}{dt} = 0$$

and

$$\sum_{in} \frac{c_n A}{v} = \sum_{out} \frac{c_n A}{v} \tag{9.12}$$

or

$$\sum_{in} \rho c_n A = \sum_{out} \rho c_n A. \tag{9.13}$$

In fluid mechanics, Equation (9.13) is referred to as the continuity equation.

EXAMPLE 9.2

Case study: changing mass of water in a boiler

A boiler is a vessel in which water is boiled to produce steam. The heating effect is often provided by the combustion of a fuel. In a typical industrial situation the demand for steam would be unsteady and the boiler control system would be designed to adapt to this automatically. The burner would be switched on and off to maintain the boiler pressure between pre-set upper and lower limits. The feed pump that supplies water to the boiler would also operate intermittently to maintain the water level in the boiler between pre-set upper and lower levels. At a given moment in time, the rate of change of mass of water substance in the boiler could be positive, negative or zero.

Feed water enters a boiler at the rate of 1.5 kg/s. Dry saturated steam leaves with a velocity of 17 m/s in a pipe of 40 mm diameter at a pressure of 6 MPa. What is the rate of change of the mass of water substance within the boiler?

SOLUTION

At 6 MPa

$$v_g = 0.032\ 44\ \mathrm{m^3/kg}$$

$$\dot{m}_{out} = c_{out} A_{out} / v_{out}$$

$$= 17 \frac{\pi \times 0.04^2}{4} \frac{1}{0.032\ 44} [\mathrm{m\ s^{-1}}][\mathrm{m^2}][\mathrm{kg\ m^{-3}}] = 0.659\ \mathrm{kg/s}.$$

Answer

$$\therefore \frac{\mathrm{d}m_{\text{sys}}}{\mathrm{d}t} = \dot{m}_{\text{in}} - \dot{m}_{\text{out}} = 1.5 - 0.659 \text{ kg/s} = 0.841 \text{ kg/s}.$$

9.5 Summary

The mass balance equation, which is the mathematical expression of the principle of conservation of mass, has been examined in this chapter. Some useful equations for mass conservation have been presented and explained.

The student should be able to

- *define*
  - a steady flow rate, an unsteady flow rate, the term *steady*, the term *non-steady* (or *unsteady*), a steady state, a steady system, a uniform velocity, a uniform flow
- *write*
  - the mass balance equation for a general system with unsteady inlet and exit flows
- *derive*
  - the expression for mass flow rate in terms of velocity, area and specific volume for uniform flow at a boundary
- *apply*
  - the principle of conservation of mass to closed systems and open systems.

9.6 Self-assessment questions

9.1 Four kilograms of ice and 10 kg of liquid water are introduced into an evacuated container that has a volume of 0.85 m^3. The sealed container is allowed to reach equilibrium at ambient temperature, which is 20 °C. What is the volume of the liquid water in the container at this temperature?

9.2 Saturated steam at a pressure of 0.7 MPa enters a steam heater. The condensate produced leaves through a device known as a steam trap, which operates intermittently. What is the rate of change of mass within the steam heater at an instant when the mass flow rate of steam into the heater is 0.0231 kg/s and the volume flow rate of saturated water from the steam heater is 0.0712 L/s?

9.3 Figure 9.3 shows the cylinder of a two-stroke diesel engine. At a given

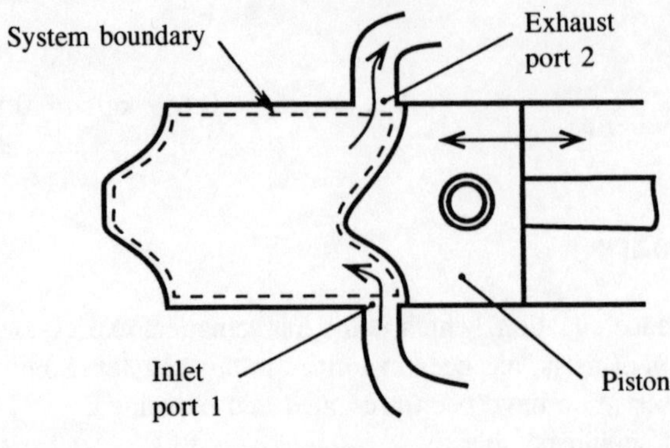

Figure 9.3

instant in time, air enters the system at a pressure of 1.1 bar absolute and a temperature of 31 °C. The inlet mass flow rate is 0.053 kg/s. Simultaneously, exhaust products leave the system at the exhaust port with a volume flow rate that equals 95% of the inlet volume flow rate. There are no other mass flows into or out of the system. The exhaust stream has the same pressure as the inlet stream but has a temperature of 400 °C. What is the rate of change of the mass within the system? As an approximation, the exhaust products may be taken to be air and ideal gas behaviour may be assumed.

CHAPTER 10

The first law of thermodynamics

In this chapter the first law of thermodynamics is explained as a fundamental relationship between heat and work for any cycle undergone by a system that has no other net effects on its surroundings. It is then used to define a difference in the property internal energy between thermodynamic states of a closed system. The non-flow energy equation is explained. Friction and fluid friction are examined in the light of the first law. Also, the first law is used to show that the adiabatic index for an ideal gas is equal to the ratio of the specific heats.

The realization that heat and work are both forms of energy transfer came rather late in the development of the science of thermodynamics. Different measurement units had been used for these two quantities. Eventually it was discovered that there was a constant of proportionality between these different units. It was possible to determine experimentally a 'mechanical equivalent of heat'. In the SI system, heat and work are expressed in the same units, the joule.

10.1 The first law of thermodynamics

The first law of thermodynamics states that the algebraic sum of the amounts of heat transfer and the amounts of work equals zero for any cycle undergone by a system that has no other net effects on its surroundings:

$$\oint Q + \oint W = 0. \qquad (10.1)$$

The circles on the summation symbols in Equation (10.1) indicate that these are summations for all the processes of a cycle.

Consider a cycle as shown in Figure 10.1. The first law of thermodynamics takes the form

$$Q_{1-2} + Q_{2-3} + Q_{3-1} + W_{1-2} + W_{2-3} + W_{3-1} = 0 \qquad (10.2)$$

where the symbols Q and W with their identifying subscripts represent the amounts of heat transfer and work for each of the processes making up the cycle.

For a cycle that is divided up into such a large number of processes that each process can involve only a small amount of heat transfer and work the first law can

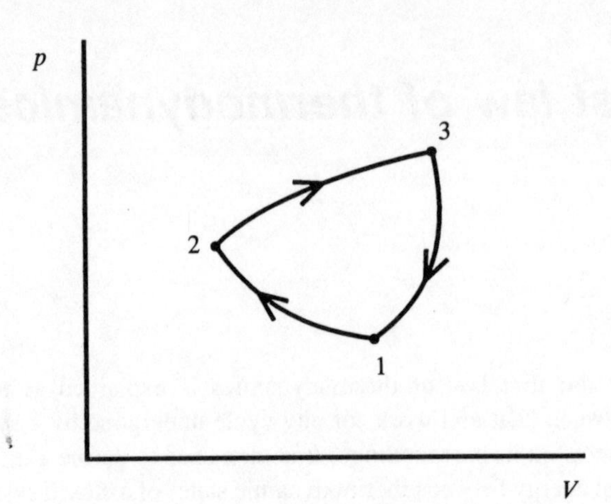

Figure 10.1 *A cycle made up of three processes.*

be written as

$$\sum \delta Q + \sum \delta W = 0 \tag{10.3}$$

where δQ and δW represent the small amounts of heat transfer and work for the many processes that make up the cycle. For a totally general cycle, which can always be considered as a series of infinitesimal processes, the first law can be written as

$$\oint \mathrm{d}Q + \oint \mathrm{d}W = 0 \tag{10.4}$$

or

$$\oint \mathrm{d}Q = -\oint \mathrm{d}W \tag{10.5}$$

The circles on the integration symbols in Equations (10.4) and (10.5) indicate that these are cyclic integrals. The infinitesimal heat transfer and work quantities must be summed up over the boundary of the system and over the period of time in which the cycle occurs.

One important consequence of the first law is that if heat and work are initially measured in different units, the conversion factor between these units can be established by means of an experiment in which a system is brought through a cycle and the net work and the net heat transfer are measured in their respective units.

The first law of themodynamics applies whenever a system whose only net effects on its surroundings are heat and work completes a cycle, i.e. returns to its initial state having undergone changes in its state. There is no requirement for the system to pass through equilibrium states. The first law has been firmly established by numerous observations and experiments. For instance, James Prescott Joule carried out a classic experiment in which stirring work was done on a closed system that contained liquid water, causing its temperature to increase. The system returned to its original state by rejecting heat to the surroundings. The heat transfer and the work were measured separately in different units and were found always to be in a fixed ratio to each other.

10.2 Internal energy and the non-flow energy equation

Another important consequence of the first law of thermodynamics is the definition of a difference in internal energy between two thermodynamic states of a closed system. Consider three processes, A, B and C, of a closed system as shown in Figure 10.2. These need not be equilibrium processes.

For cycle AC

$$Q_{1\to2,\mathrm{A}} + Q_{2\to1,\mathrm{C}} + W_{1\to2,\mathrm{A}} + W_{2\to1,\mathrm{C}} = 0. \tag{10.6}$$

For cycle BC

$$Q_{1\to2,\mathrm{B}} + Q_{2\to1,\mathrm{C}} + W_{1\to2,\mathrm{B}} + W_{2\to1,\mathrm{C}} = 0. \tag{10.7}$$

Subtracting,

$$Q_{1\to2,\mathrm{A}} - Q_{1\to2,\mathrm{B}} + W_{1\to2,\mathrm{A}} - W_{1\to2,\mathrm{B}} = 0 \tag{10.8}$$

or

$$Q_{1\to2,\mathrm{A}} + W_{1\to2,\mathrm{A}} = Q_{1\to2,\mathrm{B}} + W_{1\to2,\mathrm{B}}. \tag{10.9}$$

Therefore, $Q_{1\to2} + W_{1\to2}$ has the same value for all processes from state 1 to

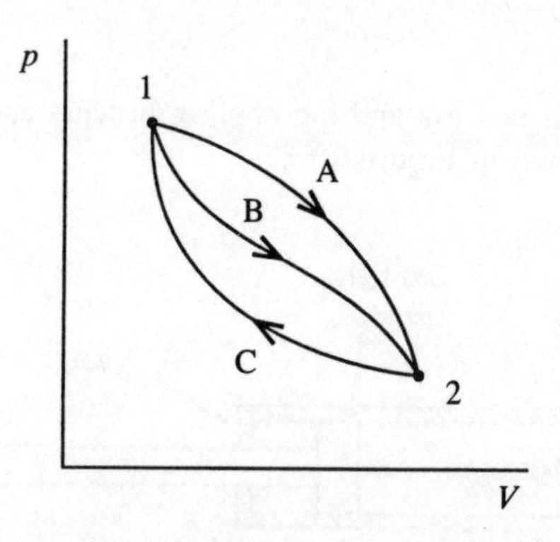

Figure 10.2 *Three processes between thermodynamic states 1 and 2 of a closed system.*

state 2. It therefore represents a difference in a property of the system between the two states. This property is given the name internal energy, U. Hence, for any non-cyclic process between states of a closed system

$$Q_{1-2} + W_{1-2} = U_2 - U_1 = \Delta U_{1-2}. \tag{10.10}$$

Equation (10.10) is known as the *non-flow energy equation*. It can also be written per unit mass of substance in the closed system as

$$q_{1-2} + w_{1-2} = \Delta u_{1-2}. \tag{10.11}$$

The non-flow energy equation is a statement that energy is conserved for any process that involves a change in the thermodynamic state of a closed system. Any analysis that makes use of this equation is known as a *first-law analysis*.

EXAMPLE 10.1

Case study: a gas spring

Figure 10.3 illustrates a gas spring, which includes a sealed space containing nitrogen gas. This type of spring is used to support the weight of the tailgate of a hatchback car. In the fully extended state of the gas spring the nitrogen has a volume of 0.732 litres. It has a pressure of 2 bar absolute and a temperature equal to that of the surroundings, 20 °C. A load is applied and the gas is compressed to 20% of its original volume. This occurs slowly enough that the system can be regarded as being in equilibrium during the process, yet quickly enough that there is little opportunity for heat transfer and the process can be regarded as adiabatic. If the two ends of the spring are immediately fixed in the compressed state and the gas spring is left for a long period of time, how much heat transfer would occur to the surroundings from the nitrogen in the spring? The adiabatic index for nitrogen is 1.4. The gas spring can be assumed to operate without friction.

SOLUTION

The compression process and the cooling process can be shown on a $p-V$ diagram, as shown in Figure 10.4.

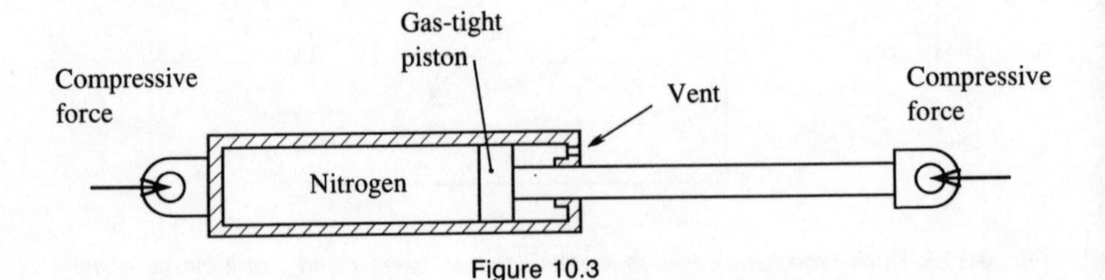

Figure 10.3

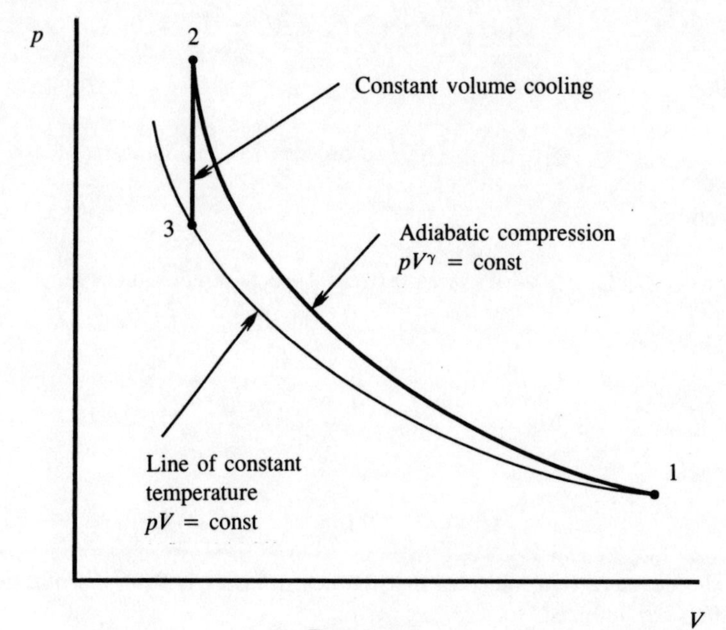

Figure 10.4

$$V_2 = 0.20V_1 = 0.20 \times 0.732 \text{ L} = 0.1464 \text{ L}$$

$$p_2 = p_1\left(\frac{V_1}{V_2}\right)^{\gamma}$$

$$= 2 \text{ [bar] } 5^{1.4} = 19.04 \text{ bar}$$

$$W_{1\to2} = \frac{p_2V_2 - p_1V_1}{\gamma - 1}$$

$$= \frac{19.04 \times 10^5 \times 0.1464 \times 10^{-3} - 2 \times 10^5 \times 0.732 \times 10^{-3}}{1.4 - 1} [\text{N m}^{-2}][\text{m}^3]$$

$$= 330.86 \text{ J.}$$

At state 3 the nitrogen will have reached the temperature of the environment, the same temperature as at state 1.

From Joule's law

$$U_3 = U_1.$$

From the non-flow energy equation

$$Q_{1\to2\to3} + W_{1\to2\to3} = U_3 - U_1 = 0.$$

But

$$Q_{1\to2\to3} = Q_{2\to3} \text{ as process } 1\to2 \text{ is adiabatic}$$

and

$$W_{1\to2\to3} = W_{1\to2} \text{ as process } 2\to3 \text{ is at constant volume.}$$

Therefore

$$Q_{2\to3} + W_{1\to2} = 0$$

Answer

$$Q_{2\to3} = -W_{1\to2} = -330.86 \text{ J.}$$

The negative sign indicates that the heat transfer is from the nitrogen to the surroundings.

10.3 First-law analysis of a friction work interaction

Whenever a friction work interaction occurs, as in Figure 10.5, which is the same as Figure 3.13, Chapter 3, the non-flow energy equation applies to the composite of the two systems that are involved. The work done by system B, which can be described as friction work, becomes internal energy of systems A and B in the region of the plane of contact. Owing to the increase in internal energy, the local temperature in this region is increased. While the friction work is occurring, and for some time afterwards, the systems will not be in thermal equilibrium. As long as there is a temperature difference within either system, or between the two, internal energy will be redistributed within them as heat transfer. If there is no heat transfer to or from the surroundings of the two systems, the internal energy of systems A and B together increases by an amount equal to the friction work. The non-flow energy equation gives

$$\Delta U_{A+B} = W_{fric} = F_B s_B. \tag{10.12}$$

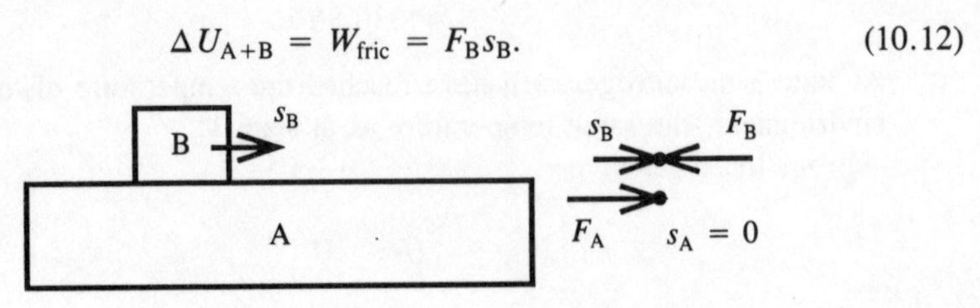

Figure 10.5 *Friction work done by a moving system,* B.

10.4 First-law analysis of fluid friction

Stirring work is mechanical shear work that is entirely dissipated by causing relative movement within a fluid. An everyday example would be stirring a cup of tea with a spoon. An engineering example would be a water brake used in engine testing. In this case the mechanical power produced by the engine is absorbed by causing a rotor to churn water. The relative movement of the rotor and the fluid gives rise to fluid friction. There is also a temporary transfer of kinetic energy to the fluid. This kinetic energy may be highly disordered, involving eddy formation and turbulent mixing of the fluid. Over time the disordered kinetic energy at the macroscopic level is redistributed down to the molecular level and so becomes internal energy. The non-flow energy equation can be applied.

EXAMPLE 10.2

By means of a churn, Figure 10.6, 195 kJ of work is done on 3 kg of water, which is initially at a temperature of 18 °C and a pressure of 1 bar. Heat transfer also occurs. What is the net heat transfer for the process if the water reaches a final equilibrium temperature of 25 °C at the same pressure?

SOLUTION

$$Q_{1-2} + W_{1-2} = U_2 - U_1 = \Delta U_{1-2}.$$

At 18 °C and 1 bar

$$u_1 \approx u_f \text{ at } 18\ ^\circ\text{C}.$$

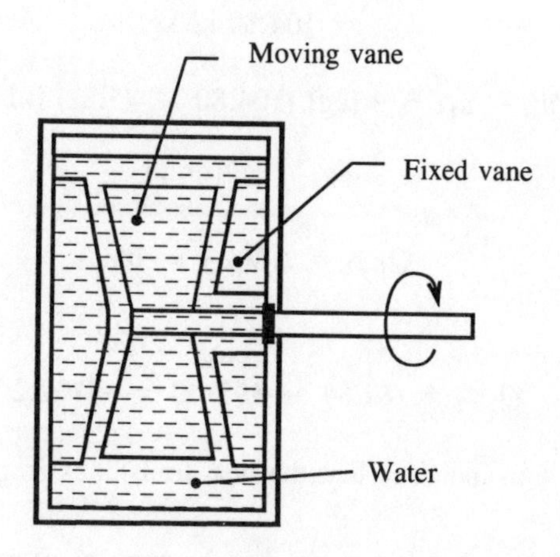

Figure 10.6 *Churn.*

At 10 °C,

$$u_f = 42.0 \text{ kJ/kg}.$$

At 20 °C,

$$u_f = 83.9 \text{ kJ/kg}.$$

Therefore, at 18 °C,

$$u_1 \approx u_f = \left[42.0 + \frac{18 - 10}{20 - 10}(83.9 - 42.0)\right] \text{ kJ/kg}$$

$$= 75.52 \text{ kJ/kg}.$$

At 25 °C and 1 bar

$$u_2 \approx u_f \text{ at } 25 \text{ °C}.$$

At 30 °C,

$$u_f = 125.7 \text{ kJ/kg}.$$

Therefore, at 25 °C,

$$u_2 \approx u_f = \left[83.9 + \frac{25 - 20}{30 - 20}(125.7 - 83.9)\right] \text{ kJ/kg}$$

$$= 104.80 \text{ kJ/kg}.$$

$$\Delta U_{1\to2} = m(u_2 - u_1) = 3 \text{ [kg] } (104.80 - 75.52) \text{ [kJ kg}^{-1}\text{]} = 87.84 \text{ kJ}$$

$$W_{1\to2} = 195 \text{ kJ}$$

$$Q_{1\to2} = \Delta U_{1\to2} - W_{1\to2}$$

Answer

$$Q_{1\to2} = (87.84 - 195) \text{ kJ} = -107.2 \text{ kJ}.$$

The negative sign indicates that the heat transfer is from the water.

The viscometer diagram of Figure 10.7 is the same as Figure 3.14, Chapter 3. The friction work or lost shear work in this case is equal to the work input to the

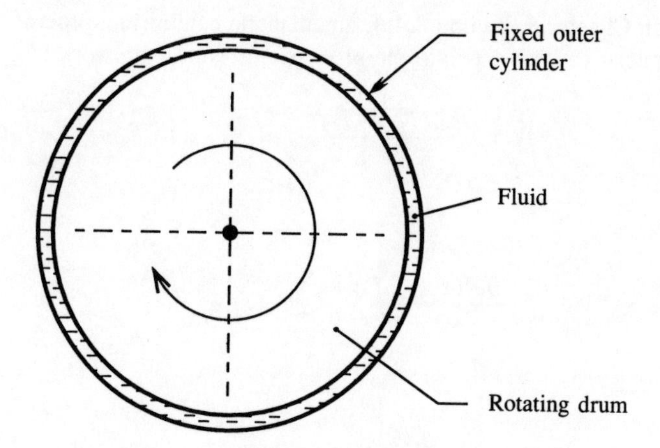

Figure 10.7 *Fluid friction within a viscometer.*

fluid from the rotating drum over a period of time, W_{fric}. If there is no heat transfer to or from the fluid, the non-flow energy equation will take the form

$$\Delta U_{1-2} = W_{fric} = \boldsymbol{T}\theta_{1-2} \tag{10.13}$$

where ΔU_{1-2} is the increase in the internal energy of the fluid for the period, $\boldsymbol{T}$ is the torque exerted on the rotating drum and θ_{1-2} is its rotation for the period.

While the friction work is occurring the fluid system is not in mechanical equilibrium. If the drum were stopped and the fluid system isolated at a particular instant, the mechanical shear forces within the fluid would change: they would decrease to zero as the fluid came to rest over a period of time. Furthermore, if the shear work loss was not uniformly distributed within the fluid volume, temperature differences would exist owing to differences in specific internal energy and the fluid system would not be in thermal equilibrium.

10.5 The adiabatic index for an ideal gas

Consider an adiabatic process of an ideal gas from equilibrium state 1 to equilibrium state 2. Assume that these states are close enough together that any change in c_v between them can be neglected.

$$Q_{1-2} = 0.$$

Therefore

$$W_{1-2} = \Delta U_{1-2} = mc_v(T_2 - T_1).$$

As stated in Chapter 8, section 8.4.3, an adiabatic equilibrium process of an ideal gas is a special case of a polytropic process, for which the work is given by

$$W = \frac{mR(T_2 - T_1)}{\gamma - 1}. \qquad \text{(8.29, Chapter 8)}$$

Therefore

$$\frac{mR(T_2 - T_1)}{\gamma - 1} = mc_v(T_2 - T_1).$$

Hence

$$R = c_v\gamma - c_v$$

and

$$\gamma = \frac{R + c_v}{c_v}.$$

But, from Equation (8.34), Chapter 8,

$$R + c_v = c_p.$$

Therefore

$$\gamma = \frac{c_p}{c_v}. \qquad \text{(8.30, Chapter 8)}$$

10.6 The significance of the first law

The first law of thermodynamics leads directly to the non-flow energy equation and embodies four important concepts, as follows:

1 The equivalence of heat and work in so far as they are both modes of energy transfer.
2 The existence of a type of energy (internal energy) that depends on the thermodynamic state of a system.
3 The possibility of measuring a difference in internal energy between thermodynamic states by making measurements of heat transfer and work.
4 The fact that energy is conserved whenever the thermodynamic state of a closed system changes.

The non-flow energy equation and the laws of mechanics (Newton's laws) constitute a major part of the principle of conservation of energy. The lost work that arises in a purely mechanical treatment of friction or fluid friction becomes internal energy.

10.7 Practical tips

- Think of the non-flow energy equation whenever you are faced with a problem involving a closed system and heat, work or energy.
- For ideal gases use the specific heat at constant volume to calculate changes in the specific internal energy. Only the initial and final temperatures are needed to find the change in the specific internal energy when the specific heat at constant volume is known.
- For substances that are not ideal gases use thermodynamic tables to look up the initial and final values of the specific internal energy.

10.8 Summary

The first law of thermodynamics has been stated in words and in mathematical form for cyclic processes of a system whose only net effects on its surroundings are heat and work. It has been shown how a difference in a property, which is known as the internal energy, can be defined from the first law. The non-flow energy equation has been introduced. First-law analyses of a friction work interaction and of a process involving fluid friction have been presented. The non-flow energy equation has been used to derive the relationship between the adiabatic index and the principal specific heats of an ideal gas. The significance of the first law has been summarized.

The student should be able to

- *state*
  — the first law of thermodynamics, four important concepts embodied in the first law
- *write*
  — the first law of thermodynamics in mathematical form, the non-flow energy equation for a closed system that undergoes a process
- *derive*
  — the non-flow energy equation from the first law of thermodynamics
- *define*
  — stirring work
- *explain*
  — the application of the non-flow energy equation to a friction work interaction and to a process involving fluid friction

- *show*
  — from the non-flow energy equation that the adiabatic index is equal to the ratio of the principal specific heats for an ideal gas

- *apply*
  — the first law of thermodynamics and the non-flow energy equation to practical problems.

10.9 Self-assessment questions

10.1 A closed rigid pressure vessel contains 0.1 m^3 of liquid water and 99.9 m^3 of dry saturated steam at a pressure of 200 kPa. The vessel is heated until the pressure is 500 kPa. Find the final temperature and the heat transfer to the water substance in the vessel.

10.2 It is found experimentally that the amount of heat transfer per unit mass required to heat a particular gas from 10 °C to 40 °C at constant pressure is 21.85 kJ/kg. Similarly, the amount of heat transfer to raise the temperature of 5.5 kg of the same gas from 7 °C to 17 °C at constant volume is found to be 28.62 kJ. Assuming that the gas is ideal and that the specific heat values are constant, calculate the values of c_p, c_v, γ and R from the measured data.

10.3 A domestic pressure cooker has a volume of 9 litres. It has been heated sufficiently to vent all air and then allowed to cool to 80 °C. It contains only 0.35 kg of water substance (liquid and vapour) at this saturation temperature. Assuming the valve remains closed, how much heat transfer to the water substance would be required to bring the pressure up to 0.15 MPa absolute?

10.4 It is not always wise, or safe, to treat liquids as having a fixed specific volume. This is particularly the case where a pressure vessel is used to store a liquid. If a rigid pressure vessel is completely filled with liquid, even quite a small temperature increase may increase the pressure dramatically and rupture the vessel. It is usual to specify the maximum permissible charge, by mass, of the pressure vessel so that this situation does not arise. The following problem illustrates how, in certain circumstances, saturated liquid can expand to fill its containment.

A steel pressure vessel contains water liquid and water vapour in equilibrium at a temperature of 20 °C. It has a volume of 0.324 m^3 and the total mass of water substance within it is 299 kg. Show that the initial specific volume of the saturated mixture is less than the specific volume at the critical point state. Sketch a p–v equilibrium phase diagram for the liquid and vapour phases of water substance and use this to show that if the temperature of the vessel is increased the volume of liquid increases and the volume of vapour decreases. At what equilibrium temperature would the entire pressure vessel become filled with liquid water? What would be the corresponding pressure? What heat transfer would be required to bring the water to this state? The steel vessel may be assumed to be rigid.

10.5 A closed system bounded by a cylinder and piston has a volume of 0.14 m^3 and contains air at a pressure of 9 bar absolute and 41 °C. The piston moves down the cylinder in such a way that the air is expanded in a polytropic process to a pressure of 2.6 bar absolute. The polytropic exponent for this process is 1.25. Evaluate the work done on the air in the cylinder and the heat transfer to it.

10.6 An ideal Stirling engine operates in a cycle as follows. Helium gas is enclosed in a sealed containment. This closed system undergoes 2900 cycles per minute. Each cycle is made up of four processes, as follows:

Process 1→2 Starting from a volume of 1.013×10^{-3} m^3, a temperature of 35 °C and a pressure of 1 MPa, the gas is compressed isothermally to one-third of its initial volume.

Process 2→3 Heat transfer occurs to the gas at constant volume until its temperature reaches 500 °C.

Process 3→4 The gas is expanded isothermally to its initial volume.

Process 4→1 Heat rejection occurs from the gas at constant volume until it reaches its original temperature of 35 °C.

(a) Sketch the form of the cycle on a $p-V$ diagram. (b) Calculate the work for each of the four processes and the net work (the algebraic sum) for the cycle. (c) Calculate the heat transfer for each of the four processes and the sum of the heat transfer quantities that are to the working fluid (i.e. positive). (d) Calculate the average net rate of work output of the gas system and the average rate of heat transfer to the gas (i.e. positive heat transfer only). For helium, c_v = 3.1159 kJ/kg K and R = 2.0772 kJ/kg K.

CHAPTER 11

The steady flow energy equation

In this chapter the steady flow energy equation is derived. Some simplifications that often arise are described. The equation is applied to a constant pressure heating or cooling process, an adiabatic work process, a throttle device and a nozzle. The concept of equilibrium in relation to steady flow systems is discussed. An equilibrium flow cycle is defined.

In a steady flow system the thermodynamic properties at any point do not change with time. Also, the shape of the system does not change and the fluid velocities are invariant with time throughout the system. Many engineering components can be regarded as steady flow open systems; for instance, a steam turbine in a power plant, a steam boiler, or even a valve.

11.1 Conservation of energy in a steady flow system

From the principle of conservation of energy (section 4.3, Chapter 4), a balance equation for the energy entering and leaving a steady flow system can be written as follows:

$$\left(\begin{array}{l}\text{rate at which}\\ \text{energy enters}\\ \text{at the system}\\ \text{boundary}\end{array}\right) = \left(\begin{array}{l}\text{rate at which}\\ \text{energy leaves}\\ \text{at the system}\\ \text{boundary}\end{array}\right). \qquad (11.1)$$

Figure 11.1 represents a general steady flow system, which is open and has just one inlet flow position and one exit flow position. Energy transfer occurs across the boundary as heat and work. Energy is also transported across the boundary by the substance that enters the system at position 1 and the substance that leaves at position 2. Equation (11.1) can be applied to this system to yield an energy balance expression in terms of the properties of the system. Two types of work are involved. One of these is shaft work and the other is normal or displacement work at the

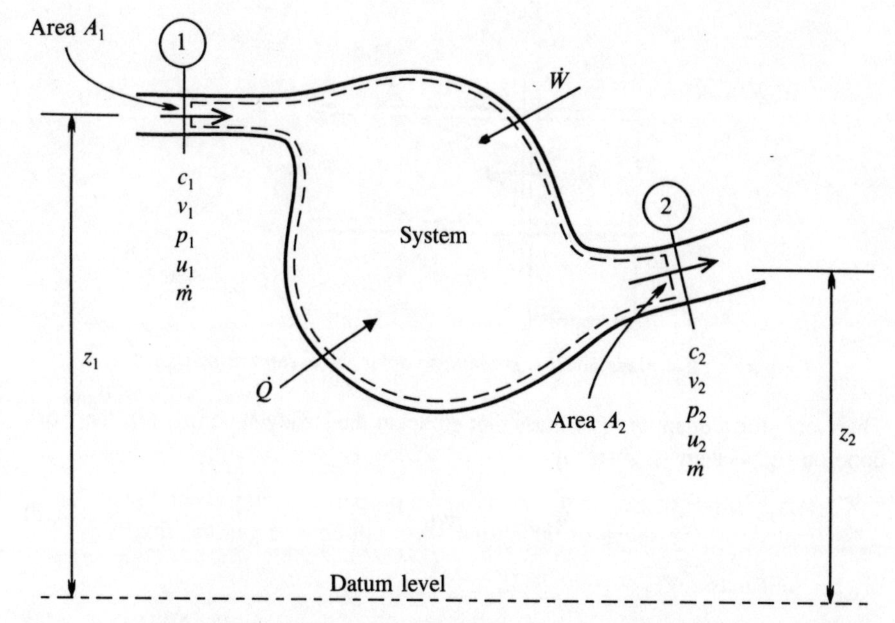

Figure 11.1 *A general steady flow fluid system with one inlet stream and one exit stream.*

positions on the boundary where the fluid enters and leaves. This is known as flow work and is described in the next section.

11.1.1 Flow work

Consider mass δm that crosses the boundary at position 1 in Figure 11.1. Figure 11.2 shows this position on a larger scale and identifies the mass at the moment before it enters the system. At position 1, a force acts on the system over area A_1. The point (or plane) of action of the force within the fluid is moving in the direction of the force, even though the system boundary is fixed. When mass δm enters, the work on the system is given by

$$\begin{aligned} \delta W_{\text{flow},1} &= \text{force} \times \text{distance} \\ &= p_1 A_1 \delta l. \end{aligned} \tag{11.2}$$

But

$$A_1 \delta l = \text{volume} = \delta m\, v_1. \tag{11.3}$$

Hence

$$\delta W_{\text{flow},1} = \delta m\, p_1 v_1. \tag{11.4}$$

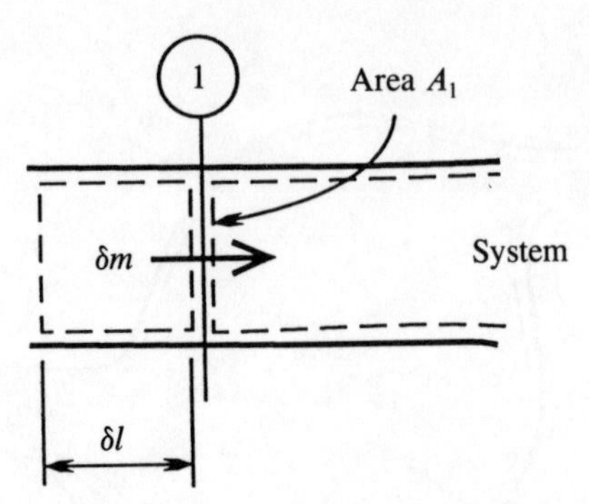

Figure 11.2 *A mass δm that is about to enter the system at position 1.*

Therefore, for a quantity m of mass that enters at the steady state (p_1, v_1), the work done on the system is given by

$$W_{\text{flow},1} = mp_1v_1 \tag{11.5}$$

or, per unit mass,

$$w_{\text{flow},1} = p_1v_1. \tag{11.6}$$

Equations (11.5) and (11.6) give the flow work and the specific flow work respectively. For a mass flow rate $\dot{m}_1$ at position 1, the rate of work input is

$$\dot{W}_{\text{flow},1} = \dot{m}p_1v_1. \tag{11.7}$$

It can be noted that the direction of the flow work is that of the mass transfer. The flow work is done on the system when the mass transfer is into the system. Flow work is done by the system whenever there is mass transfer out of the system.

11.2 The steady flow energy equation

For a steady flow system the volume and shape of the boundary cannot vary with time. Therefore, there can be no normal (or displacement) work due to volume change. The following terms describe the rates at which energy enters the system shown in Figure 11.1 at the boundary:

$\dot{Q}$ the net rate of heat transfer into the system.

$\dot{W}$ the net rate at which work, excluding flow work, is done on the system at the boundary. Generally, this term would include shaft power or electric power. There is no rate of normal (or displacement) work associated with movement of the boundary since the system is steady. Shaft power can usually be calculated from where $\boldsymbol{T}$ is torque

$$\dot{W} = \boldsymbol{T}\omega \qquad \text{(3.13, Chapter 3)}$$

and ω is the rate of rotation of a shaft.

$\dot{m}(c_1^2/2)$ kinetic energy transported into the system per unit time (assuming uniform velocity c_1 over area A_1).
$\dot{m}gz_1$ potential energy transported into the system per unit time.
$\dot{m}u_1$ internal energy transported into the system per unit time.
$\dot{m}p_1v_1$ rate at which flow work is done on the system by the surroundings at position 1.

The following terms describe the rates at which energy leaves the system at the boundary:

$\dot{m}(c_2^2/2)$ kinetic energy transported out of the system per unit time.
$\dot{m}gz_2$ potential energy transported out of the system per unit time.
$\dot{m}u_2$ internal energy transported out of the system per unit time.
$\dot{m}p_2v_2$ rate at which flow work is done by the system on the surroundings at position 2.

The *steady flow energy equation* can thus be written as

$$\dot{Q} + \dot{W} + \dot{m}\left(\frac{c_1^2}{2} + gz_1 + u_1 + p_1v_1\right) = \dot{m}\left(\frac{c_2^2}{2} + gz_2 + u_2 + p_2v_2\right). \qquad (11.8)$$

As the combination of properties $u + pv$ occurs naturally in energy equations for open systems, it is convenient to substitute the specific enthalpy, h, for this. The steady flow energy equation can therefore be written as

$$\dot{Q} + \dot{W} + \dot{m}\left(\frac{c_1^2}{2} + gz_1 + h_1\right) = \dot{m}\left(\frac{c_2^2}{2} + gz_2 + h_2\right). \qquad (11.9)$$

In many practical situations some of the terms in Equation (11.9) are negligible or cancel. The following are some particular cases:

- For an adiabatic system, $\dot{Q} = 0$.
- The work term $\dot{W}$ is often zero; for example, in a domestic hot water radiator there is no shear work done on or by the water that passes through the system.
- If $z_1 = z_2$, the potential energy terms cancel.
- If $\dot{Q} = 0$, $\dot{W} = 0$, $v =$ const, and the flow is frictionless, then (as there is no mechanism to cause a change in the temperature or specific internal energy of the fluid) $T_2 = T_1$, $u_2 = u_1$ and so

$$\frac{c_1^2}{2} + gz_1 + p_1v = \frac{c_2^2}{2} + gz_2 + p_2v. \qquad (11.10)$$

In fluid mechanics, this is known as the Bernoulli equation. If there were fluid friction there would be some conversion of kinetic energy, potential energy or flow work to internal energy and the summation of the terms on the right-hand side would be less than the summation of the terms on the left-hand side. The standard Bernoulli equation, Equation (11.10), cannot be applied where there is friction, but the steady flow energy equation, Equation (11.9), can.

EXAMPLE 11.1

Case study: a non-adiabatic turbine

Air passes through a non-adiabatic turbine at the rate of 5 kg/s and its temperature decreases by 480 K in the process. At entry to the turbine the air velocity is 150 m/s. At exit its specific volume is 1.91 m^3/kg and the flow area is 79.5×10^{-3} m^2. If the measured shaft power output from the turbine is 2.317 MW, determine the rate of heat transfer between the surroundings and the turbine and its direction.

SOLUTION

$$\dot{Q} + \dot{W} + \dot{m}\left(\frac{c_1^2}{2} + gz_1 + h_1\right) = \dot{m}\left(\frac{c_2^2}{2} + gz_2 + h_2\right)$$

$$\frac{c_1^2}{2} = \frac{150^2}{2} = 11\,250\ [(\text{m s}^{-1})^2] = 11.25\ \text{kJ/kg}$$

$$c_2 = \frac{\dot{m}v_2}{A_2} = \frac{5 \times 1.91}{79.5 \times 10^{-3}}\ [\text{kg s}^{-1}][\text{m}^3\ \text{kg}^{-1}][\text{m}^{-2}]$$

$$= 120.1\ \text{m/s}$$

$$\frac{c_2^2}{2} = \frac{120.1^2}{2}\ \text{J/kg} = 7.21\ \text{kJ/kg}.$$

Assuming air to be an ideal gas with a constant specific heat at constant pressure, from Appendix B, $c_p = 1.005$ kJ/kg K.

$$h_1 - h_2 = c_p(T_1 - T_2) = 1.005 \times 480\ [\text{kJ kg}^{-1}\ \text{K}^{-1}][\text{K}]$$

$$= 482.4\ \text{kJ/kg}$$

$$\dot{Q} = -\dot{W} - \dot{m}\left(\frac{c_1^2}{2} - \frac{c_2^2}{2} + h_1 - h_2\right)$$

$$= 2.317 \times 10^3\ [\text{kW}] - 5\ [\text{kg s}^{-1}]\ (11.25 - 7.21 + 482.4)\ [\text{kJ kg}^{-1}]$$

Answer

$$= -115 \text{ kW}.$$

The negative sign indicates that the heat transfer is from the turbine to the surroundings.

11.3 The constant pressure heating or cooling process

A device in which a fluid is heated or cooled in steady flow is often idealized as a system in which

there is negligible pressure drop,
the changes in kinetic and potential energy between entry and exit are negligible
and where no shear work is done on or by the system at the boundary.

Figure 11.3 is a schematic representation of a steady flow system in which heat transfer occurs at constant pressure. Some examples would be

a boiler, in which liquid water enters and steam leaves,
a condenser, in which a vapour enters and condensed liquid leaves,
or either side of a liquid-to-liquid heat exchanger, where a liquid is heated or cooled from one temperature to another.

The steady flow energy equation reduces to

$$\dot{Q} = \dot{m}(h_2 - h_1). \tag{11.11}$$

The rate of heat transfer can be calculated if the mass flow rate and the specific enthalpy values at entry to and exit from the system are known. If the fluid does not change phase and has a constant specific heat, then

$$\dot{Q} = \dot{m}c_p(T_2 - T_1) \tag{11.12}$$

$$= \dot{m}c_p(t_2 - t_1). \tag{11.13}$$

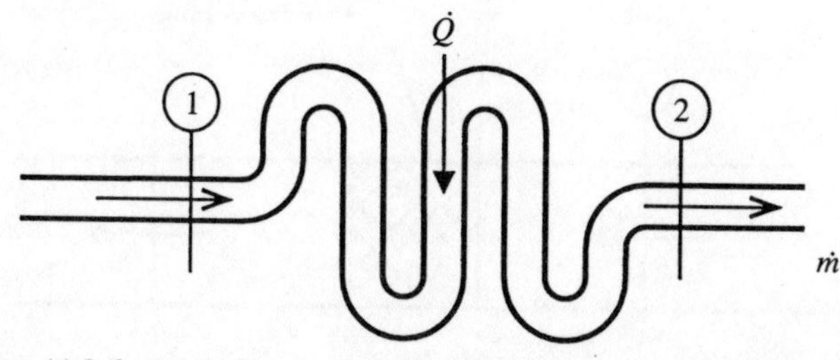

Figure 11.3 *A steady flow heat transfer process at constant pressure.*

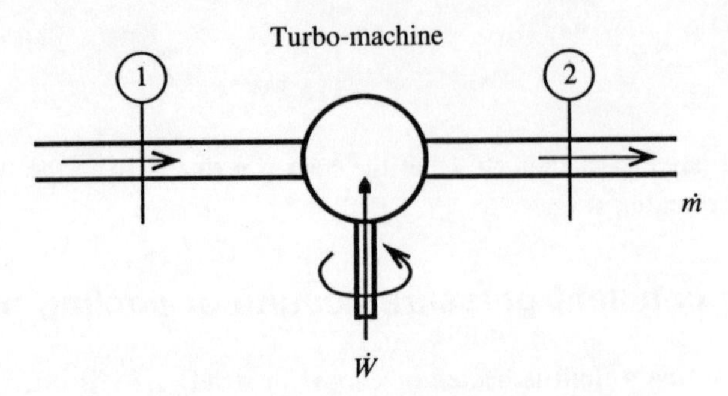

Figure 11.4 *A steady flow, adiabatic work process.*

11.4 The adiabatic work process

Devices such as pumps, turbines, fans and compressors are often idealized as being steady flow devices where the heat transfer is negligible (in comparison with the work) and changes in the potential energy and kinetic energy between entry and exit are negligible. Figure 11.4 is a schematic diagram of this type of device. The steady flow energy equation reduces to the form

$$\dot{W} = \dot{m}(h_2 - h_1). \tag{11.14}$$

11.5 The adiabatic throttling process

A *throttling process* is a flow process in which a restriction such as a porous plug or a partially open valve causes a pressure drop. Figure 11.5 illustrates a throttling process. As throttling usually occurs within a system boundary that has a small surface area, heat transfer is usually negligible. Also, any change in elevation is usually negligible, so $z_1 \approx z_2$. The kinetic energy terms in the steady flow energy equation can usually be omitted too on the basis that the velocity after throttling is about the same as that before throttling (in which case $c_1^2/2 \approx c_2^2/2$) or because both velocities

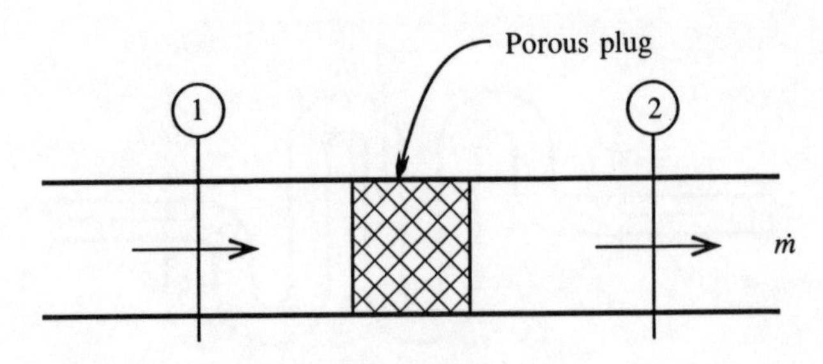

Figure 11.5 *A throttling process.*

are low enough that both kinetic energy terms are negligible (so that $c_1^2/2 \approx 0 \approx c_2^2/2$). No shear work is done on or by the system at the boundary and so, from the steady flow energy equation,

$$h_1 = h_2. \tag{11.15}$$

Therefore, the enthalpy downstream of a throttling process is the same as the enthalpy upstream.

EXAMPLE 11.2

Case study: steam pressure reduction in a throttle valve

Steam with a dryness fraction of 0.97 passes through a throttle valve where its pressure is reduced from 2.0 MPa to 0.1 MPa. What is the temperature of the steam before and after the valve?

SOLUTION

From the steam tables at 2.0 MPa, $h_f = 908.6$ kJ/kg, $h_{fg} = 1888.6$ kJ/kg and $t_s = 212.4$ °C.

Answer

Therefore, the temperature of the steam before the valve is 212.4 °C.

$$h_1 = h_f + xh_{fg}$$

$$h_1 = 908.6 \text{ kJ/kg} + 0.97 \times 1888.6 \text{ kJ/kg}$$

$$h_1 = 2740.5 \text{ kJ/kg}.$$

For the throttling process

$$h_2 = h_1 = 2740.5 \text{ kJ/kg}.$$

From the steam tables at 0.1 MPa, $h_g = 2675.4$ kJ/kg. As h_2 is greater than this, the steam leaving the valve is superheated. Also, at this pressure $t_s = 99.6$ °C. The entry and exit states are shown in Figure 11.6.

At 0.1 MPa and 150 °C, $h = 2776.3$ kJ/kg. Therefore,

$$t_2 = 99.6 + \frac{2740.5 - 2675.4}{2776.3 - 2675.4}(150 - 99.6) \text{ °C}$$

Answer

$$= 132.1 \text{ °C}.$$

This is the temperature after the valve.

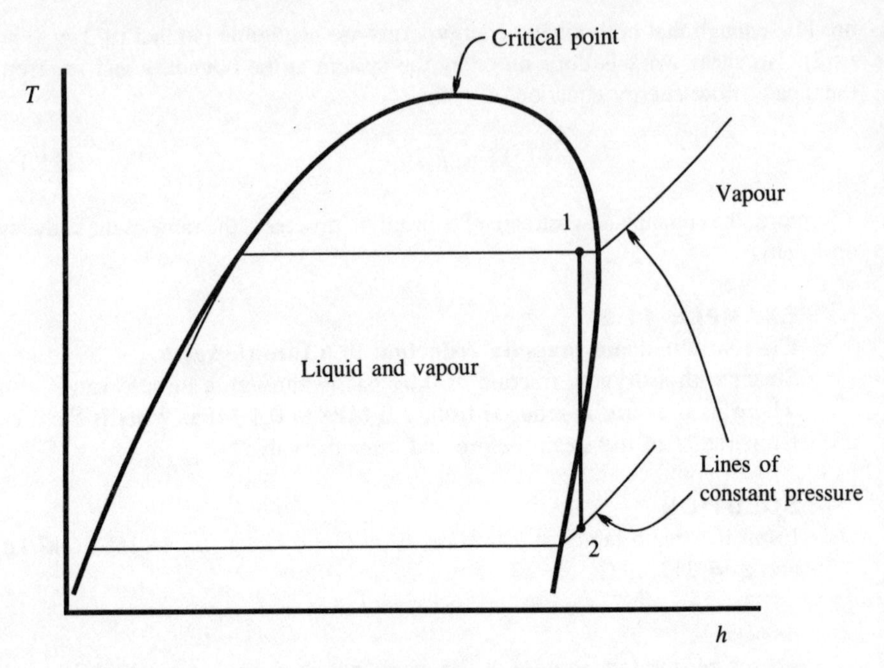

Figure 11.6

11.6 The adiabatic nozzle

A *nozzle* is a device in which a fluid is accelerated owing to a pressure drop. Heat transfer is usually negligible owing to the relatively small surface area of the nozzle and the relatively short transit time of the fluid. Figure 11.7 illustrates a nozzle.

For a nozzle, $\dot{W} = 0$, $\dot{Q} = 0$, $z_1 = z_2$, and hence

$$\frac{c_1^2}{2} + h_1 = \frac{c_2^2}{2} + h_2. \tag{11.16}$$

If, in addition, $c_1^2/2$ is negligible, then

$$c_2 = \sqrt{2(h_1 - h_2)}. \tag{11.17}$$

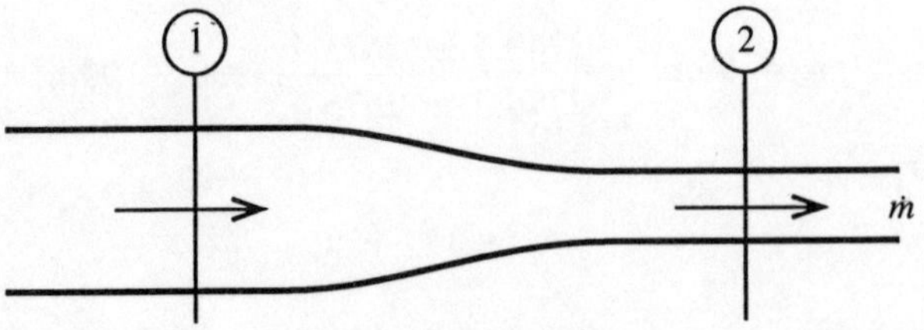

Figure 11.7 *A nozzle.*

EXAMPLE 11.3

Case study: gas flow in a stationary blade passage within an aircraft engine

Figure 11.8 is a schematic representation of a system boundary for the analysis of the flow between two adjacent stationary blades within an aircraft jet engine. In the turbine expander of the engine there are rings of fixed blades like these that direct the gas onto similar rings of moving blades that are attached to a rotating shaft. The boundary in Figure 11.8 has been chosen so that the air enters normal to its surface at position 1 and leaves normal to its surface at position 2. There is no flow across any other part of the boundary. The gas at position 1 has a pressure of 4.8 bar, a temperature of 360 °C and negligible velocity. It contains some combustion products, but its properties can be taken to be the same as those of pure air. It undergoes an adiabatic equilibrium process within the blade passage and leaves at position 2 with a pressure of 4.1 bar. Calculate the exit temperature and the exit velocity. This analysis would form part of the design calculations for the engine.

SOLUTION

From Equation (8.25) for an adiabatic equilibrium process of an ideal gas

$$\frac{T_2}{T_1} = \left(\frac{p_2}{p_1}\right)^{(\gamma-1)/\gamma}.$$

$$T_2 = (360 + 273.15)\left(\frac{4.1}{4.8}\right)^{(1.4-1)/1.4} \text{ K}$$

$$= 605.27 \text{ K}$$

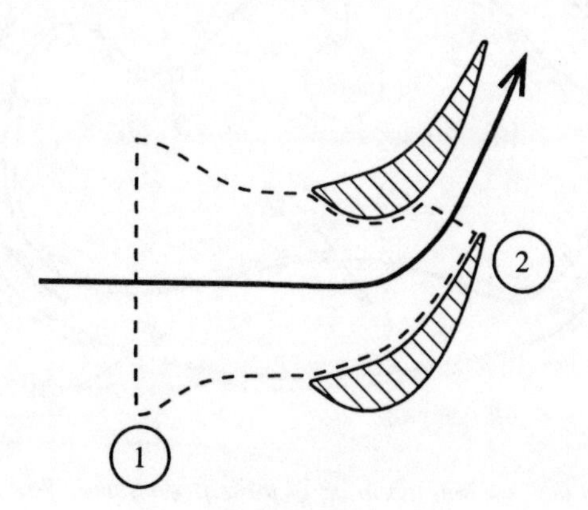

Figure 11.8

Answer

$$t_2 = 605.27\ [\mathrm{K}] - 273.15\ [\mathrm{K}] = 332.12\ ^\circ\mathrm{C}.$$

$$C_2 = \sqrt{2(h_1 - h_2)} = \sqrt{2c_p(T_1 - T_2)} = \sqrt{2c_p(t_1 - t_2)}$$

$$= \sqrt{2 \times 1.005 \times 10^3(360 - 332.12)\ [\mathrm{J\ kg^{-1}\ K^{-1}}][\mathrm{K}]}$$

Answer

$$= 236.7\ \mathrm{m/s}.$$

11.7 Flow cycles

A *closed flow system* is a closed system that can be regarded as made up of linked open flow systems. Figure 11.9 shows such a system, which comprises three linked flow systems.

A *flow cycle* is a cycle in which matter undergoes flow processes as it passes around a closed flow system. The flow cycle can be visualized as the series of processes undergone by point systems that pass around a closed flow system. The closed flow system shown in Figure 11.9 contains a flow cycle.

11.7.1 Steady flow systems and thermodynamic equilibrium

In reality, a steady flow system in which a flow process or a flow cycle occurs is not in an equilibrium thermodynamic state. This follows from the definition of an equilibrium thermodynamic state, namely: an *equilibrium thermodynamic state* is

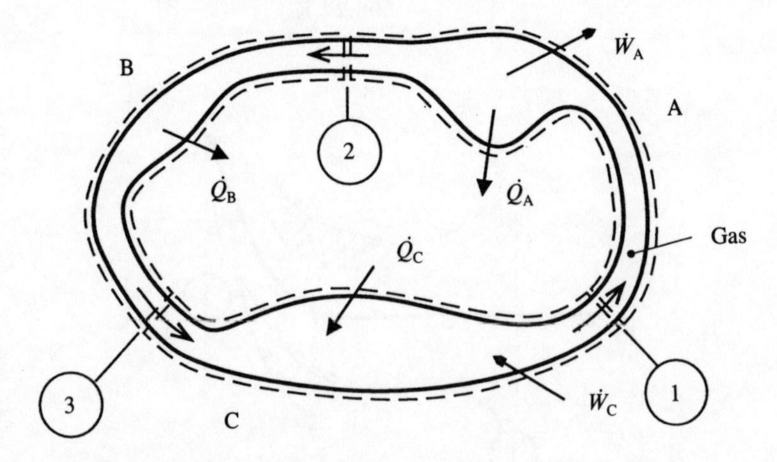

Figure 11.9 *A closed flow system made up of three open steady flow systems* A, B *and* C. *This system contains a flow cycle.*

a state such that if the system were isolated from its surroundings no change in its thermodynamic properties would occur (Chapter 2, section 2.2.2). If the system were isolated from its surroundings, a change in its thermodynamic properties would occur. It would reach an equilibrium state that could be described by, for instance, the equilibrium temperature, the equilibrium pressure and the specification of the equilibrium chemical composition of the entire system. If instead different subsystems were isolated, the parameters describing their equilibrium thermodynamic states would differ from those that would describe the isolated equilibrium thermodynamic state of the whole system.

However, it is possible to imagine a flow system containing an infinite number of point systems that all undergo the same sequence of equilibrium flow processes. These point systems would be constrained so that no heat transfer, shear work transfer, chemical reaction or exchange of substances could take place between the point systems directly. All such interactions would occur with an infinite number of external heat sources and sinks, work sources and sinks, and sources and sinks of substances. The point systems would be in equilibrium with the external sources and sinks. These constraints would be imagined to hold if any part of the flow system were isolated so that the thermodynamic states of the point systems it contained would not change. The parameters required to describe the equilibrium thermodynamic state of the steady flow system would consist of those parameters necessary to describe the whole range of equilibrium thermodynamic states it contained. If any particular subsystem were isolated, the parameters required to describe its thermodynamic state would be a subset of those required to describe the thermodynamic state of the whole system.

11.7.2 Equilibrium flow cycles

An *equilibrium flow cycle* is a flow cycle in which the point systems that undergo the processes of the cycle are considered to be in thermodynamic equilibrium at every position within the cycle. This is a theoretical ideal that can be approached,

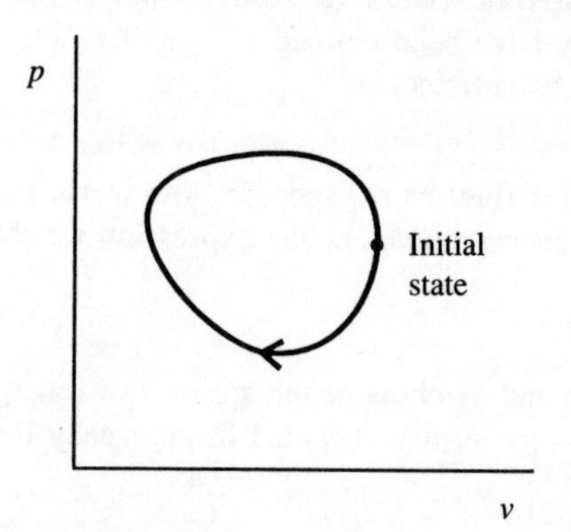

Figure 11.10 *A cycle that could be contained by a closed flow system as in Figure 11.9.*

but never realized. An equilibrium flow cycle can be represented on a state diagram of intensive properties in the same way as an equilibrium cycle of a simple closed system can be represented. Figure 11.10 shows an equilibrium cycle that could be contained by a closed steady flow system.

11.8 Practical tips

- There are many cases where the steady flow energy equation reduces to just a few terms. However, it is a good policy to consider the full equation in every case and to eliminate consciously those terms that can be left out or cancelled.
- Take care with the units of the terms in the steady flow energy equation. In thermodynamic tables, specific enthalpy values are often given in kJ/kg. However, the velocity is likely to be expressed in m/s, which would result in $c^2/2$ being in J/kg. All terms should be brought to the same units by inserting the appropriate conversion factors.
- Remember to use the steam tables for the properties of water and steam, but to use the ideal gas relationships and the specific heat values to determine the properties of gases that can be considered ideal.

11.9 Summary

Flow work has been explained. The steady flow energy equation has been derived. Some important cases where this reduces to a simpler form have been described. The relationship of the steady flow energy equation to the Bernoulli equation has been mentioned. The constant pressure heating or cooling process, the adiabatic work process, the adiabatic throttling process and the adiabatic nozzle have been analyzed. A closed flow system and a flow cycle have been defined. Thermodynamic equilibrium has been discussed in relation to steady flow systems. A flow cycle and an equilibrium flow cycle have been defined.

The student should be able to

- *derive*
  — expressions for flow work, specific flow work, rate of flow work; the steady flow energy equation; the expression for the exit velocity from a nozzle

- *explain*
  — all the terms and symbols of the steady flow energy equation, why the specific enthalpy term is included in the steady flow energy equation

- *define*
  — a throttling process, a nozzle, a closed flow system, a flow cycle, an equilibrium flow cycle

- *describe*
  - an ideal, steady flow, constant pressure heating process; an ideal, steady flow, adiabatic work process; an ideal, steady flow throttling process; an ideal, steady flow adiabatic nozzle; the relationship between the steady flow energy equation and the Bernoulli equation
- *sketch*
  - a general steady flow fluid system with one inlet and one outlet flow position with appropriate symbols for the derivation of the steady flow energy equation, a diagram to help explain flow work at a system boundary
  - a steady flow heating or cooling device, a steady flow work device, a steady flow throttling device, a steady flow nozzle, a closed flow system made up of steady flow open systems
- *apply*
  - the steady flow energy equation to practical problems while making appropriate assumptions about which terms, if any, can be neglected
- *discuss*
  - the concept of thermodynamic equilibrium in relation to steady flow systems.

11.10 Self-assessment questions

11.1 The feed water entering a boiler has a temperature of 50 °C. Steam leaves the boiler at 1.2 MPa with a dryness fraction of 96%. The mass flow rate of the steam is 1396 kg/h. The mass flow rate of the feed water has the same value. Determine the net rate of heat transfer to the water substance in the boiler.

11.2 Water enters a steady flow system at the rate of 7.23 kg/s. It has a temperature of 25 °C, a pressure of 1.5 MPa and a velocity of 75 m/s. The only other flow through the system boundary is steam, which leaves the system at the same pressure. It has a dryness of 96%, a velocity of 107 m/s and is at an elevation of 134 m with respect to the inlet level of the water. The system produces a net shaft work output at the rate of 2.25 MW. Determine the net rate of heat transfer across the boundary and state its direction.

11.3 A steady flow rate of 0.26 kg/s of air passes through a compressor. The air enters with negligible velocity at a temperature of 21 °C. It leaves the compressor at 95 °C with negligible velocity. Evaluate the increase in specific enthalpy between the inlet and exit states. If the rate of work input to the compressor is 19.78 kW, calculate the rate and the direction of the heat transfer between the surroundings and the compressor.

11.4 Steam with a mass flow rate of 1.5 kg/s enters a turbine with a pressure of 3 MPa and a temperature of 350 °C. It leaves with a pressure of 0.15 MPa and a dryness fraction of 97%. Work output is produced at the rate of 736.5 kW. Is this an adiabatic turbine?

CHAPTER 12

Heat engines

The concept of a heat engine is introduced in this chapter. Many practical applications of thermodynamics involve engines for producing work, or engines operating in reverse for producing cooling or heating effects. The thermal efficiency and the coefficient of performance are introduced as dimensionless ratios for quantifying the thermodynamic performance of engines and engines that operate in reverse respectively.

The expression 'heat engine' (or 'thermodynamic engine') in thermodynamics has a precise meaning that is somewhat more restricted than the word 'engine' in everyday language and in engineering generally. The precise meaning of the term 'heat engine' is described in this chapter.

12.1 Thermal reservoirs

In general, a heat source may provide and a heat sink may accept heat transfer over a range of temperatures. Also, the temperature of a heat source or sink may vary as it undergoes heat transfer interactions with other systems.

A *thermal reservoir* is a system of fixed temperature that can accept or provide heat transfer without undergoing any change in its own temperature.

12.2 Work reservoirs

A *work reservoir* is a system that can accept or provide mechanical shaft work or its equivalent.

An *equivalent to mechanical shaft work* (or shear work) is any energy transfer that could be converted to an equivalent energy transfer as mechanical shaft work while having no other net effect on the surroundings.

An example of an equivalent to mechanical shaft work would be electric power transfer over a time interval (see section 3.3.3, Chapter 3). Normal work due to a point force is equivalent to mechanical shaft work. However, normal work due to a force distributed over an area is not, as it also involves either a displacement of the boundary (and thus a volume change) or the transfer of substance across the boundary.

When a work reservoir is specifically a reservoir of mechanical work, it can be described as a *mechanical reservoir*. The term 'work reservoir' is more appropriate than *work source* or *work sink* for a system that can both accept and provide work.

12.3 The heat engine

A *heat engine* or *thermodynamic engine* is a system that receives heat transfer from one or more heat sources, transfers heat to one or more heat sinks, provides a net work output to a work reservoir, has no other net effect on its surroundings and undergoes no net change in its state.

It is a consequence of the second law of thermodynamics, which is introduced in Chapter 13, that heat transfer to a heat engine must occur at temperatures higher than those at which heat is rejected if there is to be a net work output. This requirement is stated formally as the Clausius inequality, which is presented in section 13.9, Chapter 13. For the present, it is sufficient to keep in mind that the temperature of the heat source must be higher than that of the heat sink.

There are several ways that a system can function as a heat engine while undergoing no net change in its state. This requirement can be met when a system undergoes a cycle. It can also be met when a system contains a flow cycle, which could be steady or unsteady. An electronic system that has a heat source and a heat sink and produces an electric power output equivalent to the net rate of heat transfer is another example of a system that undergoes no net change in its state and can be regarded as a heat engine.[11]

The standard representation of a heat engine is shown in Figure 12.1(a). Figure 12.1(b) shows a heat engine where the heat source and sink are thermal reservoirs. The subscripts H and L atttached to the heat transfer symbols in Figure 12.1(b) indicate that the heat transfer interactions occur at the fixed temperatures of the thermal reservoirs, T_H and T_L, respectively.

From the first law of thermodynamics

$$\oint \mathrm{d}Q = -\oint \mathrm{d}W. \qquad \text{(10.5, Chapter 10)}$$

$$\text{(net heat transfer to system)} = -\text{(net work done on the system)} \qquad (12.1)$$

$$\text{(heat received)} - \text{(heat rejected)} = \text{(net work done by the system)}$$

$$Q_{\text{in}} - Q_{\text{out}} = W_{\text{net,out}}. \qquad (12.2)$$

11. Such systems exist. The production of an electromotive force from a temperature difference between two junctions of dissimilar metals is due to two effects, known as the Seebeck effect and the Peltier effect. If current is drawn, then electric power is produced.

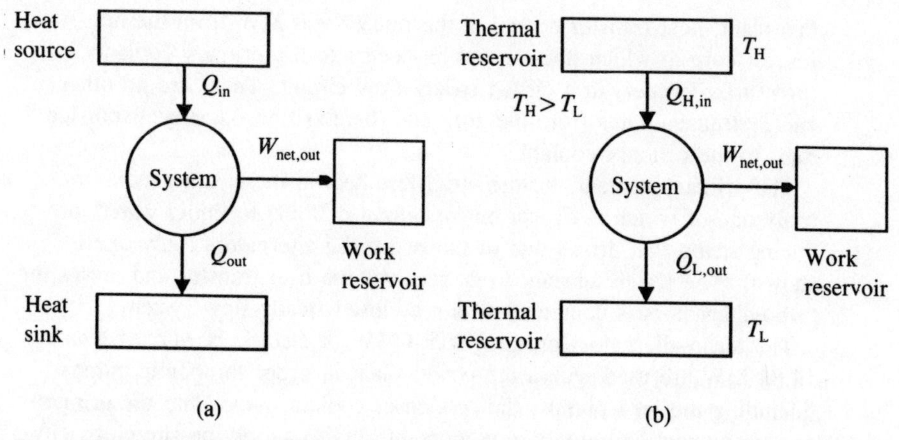

Figure 12.1 *(a) A heat engine, and (b) a heat engine where the heat source and sink are thermal reservoirs.*

12.3.1 *Thermal efficiency*

The *thermal efficiency* of a heat engine is the ratio of the net work output to the amount of heat transfer into the system:

$$\eta_{th} = \frac{W_{net,out}}{Q_{in}}. \tag{12.3}$$

Note: $W_{net,out}$ is the *net* work *output from* the system and Q_{in} is the heat transfer *into* the system.

From Equations (12.2) and (12.3), the thermal efficiency can also be written as

$$\eta_{th} = \frac{Q_{in} - Q_{out}}{Q_{in}} \tag{12.4}$$

$$= 1 - \frac{Q_{out}}{Q_{in}}. \tag{12.5}$$

It can be concluded from Equation (12.5) that the thermal efficiency of a heat engine can only be 100% if there is no heat rejection from the engine. It will be seen in Chapter 13, section 13.9.1, that the second law of thermodynamics specifically excludes this possibility.

EXAMPLE 12.1

Case study: nuclear power plant

Figure 12.2 is a schematic representation of a nuclear power plant. In this plant, heat transfer occurs at the rate of 900 MW from the hot reactor core in which nuclear fission occurs to the primary coolant (pressurized water) in a closed steady flow circuit. There are no other energy transfer rates from the core and there are no other heat transfer rates to the primary coolant.

The primary coolant, in turn, transfers heat in the steam generators (only one of which is shown out of a total of four) to liquid water, producing steam that drives one of the two turbo-alternators (only one is shown). The water substance that receives the heat transfer and drives the turbo-alternators is contained within a closed steady flow system.

The turbo-alternators produce 305.3 MW of electric power, of which 14.88 MW are used within the power plant in order to operate pumps (including the feed pumps, the condenser coolant pumps and the primary coolant pumps) and auxiliary equipment, and to supply on-site electricity needs. The balance of the electric power is supplied to the external electric network, the national grid. No component of the plant other than the turbo-alternators produces a work output. The steam that leaves the turbo-alternators is condensed in the condenser, transferring heat to sea water. The condensate is recirculated, via the feed pumps, to the steam generators.

The plant operates in a steady state. From various parts of the plant there is heat rejection to the atmosphere. This heat rejection is small in comparison with that to the sea water in the condenser. Figure 12.3 is a

Table 12.1

System	*Description*
A	Reactor core (fuel rods and control rods). This provides heat transfer to the primary coolant
B	Primary coolant circuit (compressed liquid water)
C	Water and steam circuit (liquid water and steam)
D	Turbo-alternator
E	Feed pumps and driving motors
F	Primary coolant pumps and motors
G	Condenser coolant pumps and motors
H	Condenser coolant system (sea water). The sea water flows through this system
I	Electric junction. This receives electric power from the turbo-alternators and supplies power to the pumps and auxiliary equipment and to the national grid.
J	Auxiliary equipment and on-site electric power consumption
K	The sea (sea water). This provides cold sea water as a coolant and accepts back warm sea water
L	The atmosphere (air). This receives heat transfer from various parts of the plant
M	The national grid. This serves as the sink of electric power

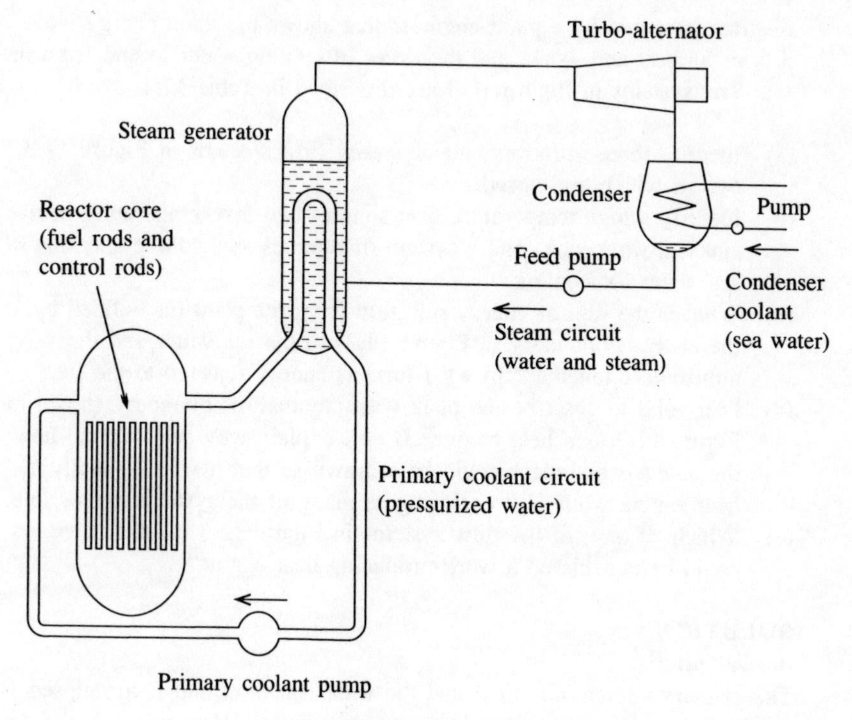

Figure 12.2

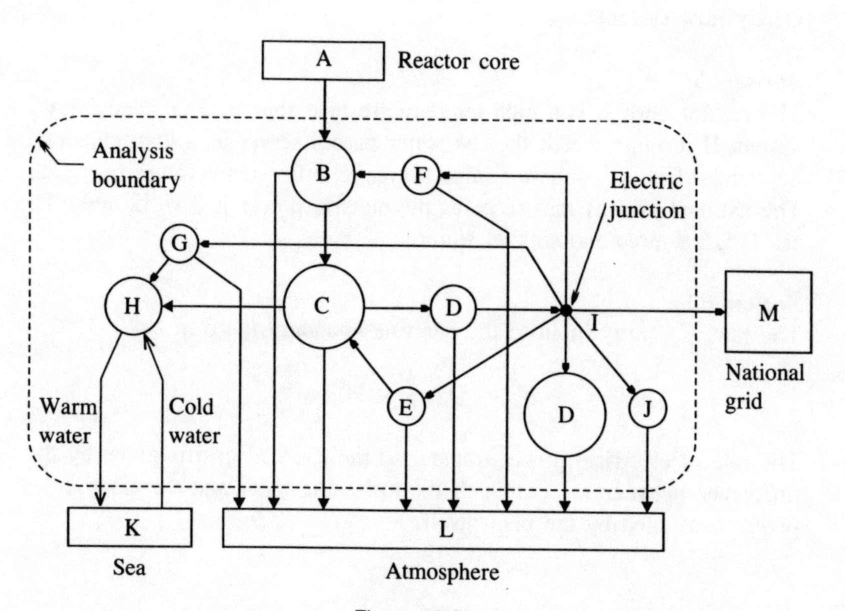

Figure 12.3

diagram prepared by a plant engineer that shows the main energy interactions as heat and work, and the flows of cooling water to and from the sea. The systems in Figure 12.3 are described in Table 12.1.

(a) Identify three, different, major steady flow systems in Figure 12.3, two of which are closed.
(b) Identify a high-temperature heat source, two low-temperature heat sinks, a work sink, and a system that serves as a source and sink of sea water for cooling.
(c) What is the rate of energy rejection from the plant (as defined by the analysis boundary in Figure 12.3) to the sea water and the atmosphere together? In what form is energy rejected to the sea?
(d) Is it valid to describe the plant with the analysis boundary shown in Figure 12.3 as a heat engine? If not, explain why and indicate how the analysis boundary could be redrawn so that it would specify a heat engine while still enclosing as many of the systems as possible.
(e) Which, if any, of the flow systems in Figure 12.3 on their own could be considered a work-producing heat engine?

SOLUTION

Answer (a)

The primary coolant circuit B and the water/steam circuit C are closed steady flow systems. The condenser coolant system H in which sea water passes through the condenser coolant pump and the condenser is an open steady flow system.

Answer (b)

The reactor core A is a high-temperature heat source. The steady flow system H through which the sea water passes serves as a low-temperature heat sink. The atmosphere L also serves as a low-temperature heat sink. The national grid M that receives the electric power is a work sink. The sea K is a source and sink of water.

Answer (c)

The rate of energy input to the analysis boundary is given by

$$\dot{E}_{\text{in}} = \dot{Q}_{\text{A}\rightarrow\text{B}} = 900 \text{ MW}.$$

The rate of electrical power transfer to the national grid is given by the difference between the output of the turbo-alternator and the electric power consumed by the plant itself:

$$\dot{W}_{\text{I}\rightarrow\text{M}} = 305.3 - 14.88 \text{ MW} = 290.42 \text{ MW}.$$

Hence, as the plant is operating in a steady state, from the principle of conservation of energy the rate of energy rejection to the sea and the atmosphere is given by

$$\dot{E}_{\text{sea+atm}} = \dot{E}_{\text{in}} - \dot{W}_{\text{I}\rightarrow\text{M}}$$

$$= 900 - 290.42 \text{ MW}$$

$$= 609.58 \text{ MW}.$$

Energy is rejected to the sea at a steady rate in the form of a net rate of enthalpy transport to the sea, which could also be described as a net rate of flow work and a net rate of transport of internal energy.

Answer (d)

The analysis boundary shown in Figure 12.3 does not specify a heat engine, as the system it encloses has effects on its surroundings other than heat transfer and work: it involves a net transport of enthalpy to the sea. By redrawing the analysis boundary to exclude system H, the condenser coolant system, a heat engine would be specified. This would have the reactor core A as its heat source, the atmosphere L and the condenser coolant system H as heat sinks, and the national grid M and the condenser coolant system H as work sinks.

Alternatively, the analysis boundary could be moved out into the sea, system K, to such a distance that there would be no flow or transport of enthalpy across it. Energy would only cross this modified boundary as heat transfer and, with respect to it, there would be only one work sink, system M.

Answer (e)

Three flow systems have been identified within the plant: the primary coolant flow system B, the water/steam flow system C and the condenser coolant system H. The primary coolant system B receives heat transfer from a heat source and rejects heat to a heat sink at a lower temperature. However, the net work is an input (from the primary coolant pumps), not an output, and on this basis the flow system does not qualify as a work-producing heat engine.

The water and steam system C receives heat transfer from system B. It rejects heat to systems H and L. There is a work output rate to the turbines (of greater than 305.3 MW, as the turbo-alternators will not be perfectly efficient). There is a work input rate from the feed pumps (which is some fraction of the 14.88 MW of power used internally in the plant). Therefore, the net work rate is an output. It is equivalent to the net heat input rate since the system is closed (there is no transport of energy across its boundary) and is in a steady state. The system undergoes no change in its state. Therefore, it can be considered a work-producing heat engine.

The steady flow system H, through which the sea water passes, does not meet the criteria for it to be considered a work-producing heat engine. There is no heat sink, the net work is an input and there is a net transport of enthalpy to the surroundings (which is an effect that does not qualify as either heat or work).

12.4 Combustion engines

If a fuel is burned in air in a steady flow system, this system can serve as a heat source for a heat engine. A fuel-fired steam power plant that produces electricity can be regarded as a heat engine that receives heat transfer from a steady flow combustion system. Such a plant can be described as an engine with external combustion. However, if the system boundary is drawn to include the entire plant, including the combustion process, as shown in Figure 12.4, there are two sources of matter and one sink of matter. There is one heat sink and one work sink. The net work output is not equivalent to the *net* heat input, as there is a large net transport of enthalpy into the system. This does not satisfy the definition of a heat engine as it has effects on its surroundings other than heat transfer and work and as it does not receive heat transfer from a heat source.

A *combustion engine* is a system in which a fuel and air or oxygen are taken in, the fuel is burned, combustion products are rejected, heat is rejected to a heat sink and there is a net work output while the system undergoes no net change in its state.

An *internal combustion engine* is a combustion engine in which the air or oxygen for combustion, the fuel (possibly)[12] and the combustion products are directly involved in the main work processes.

A combustion engine can function while undergoing no net change in two ways. It can undergo a cycle made up of unsteady processes, returning to its initial state after a period of time or periodically. In this case, the cycle includes the replacement of the spent charge with a fresh charge of fuel and oxidant. A spark ignition reciprocating engine is of this type. Alternatively, a combustion engine may be in a steady state as the working fluids pass through it (and combustion takes place) in steady flow producing a net power output. An internal combustion gas turbine engine is of this type.

Internal combustion engines such as the spark ignition engine used in cars and the gas turbine engine used in aircraft do not normally satisfy the strict requirements for a system to be considered a heat engine. They have net effects on their surroundings other than heat and work interactions: they accept air and fuel, and discharge

12. In a compression ignition engine the fuel is injected into the cylinder at the end of one of the main work processes, the compression process.

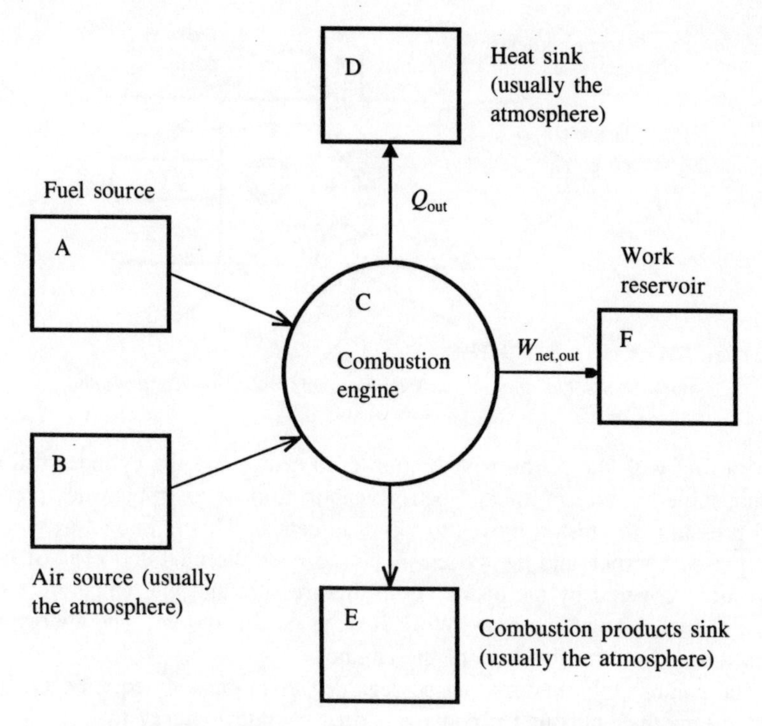

Figure 12.4 *A general schematic diagram of a combustion engine.*

hot combustion products. However, for each type of internal combustion engine, it is possible to define either a non-flow cycle or a steady flow cycle that is similar to the actual cycle and that can be considered a heat engine. The thermal efficiency of this closely related cycle can be calculated and the factors that limit its value can be investigated. This is often a useful simplification for a first analysis. It can be followed, if necessary, by a full thermodynamic analysis of the actual cycle.

12.4.1 The processes of a two-stroke, spark ignition, internal combustion engine

Figure 12.5 is a schematic representation of a two-stroke, spark ignition, internal combustion engine. The piston, which is linked by a connecting rod to the engine crankshaft, moves with a reciprocating motion in the cylinder.

The *top-dead-centre position* is the position when the piston is nearest the cylinder head (containing the spark plug).

The *bottom-dead-centre position* is the position when the piston is furthest from the cylinder head.

The processes that occur in the cylinder, as shown in Figure 12.6(a), are the following:

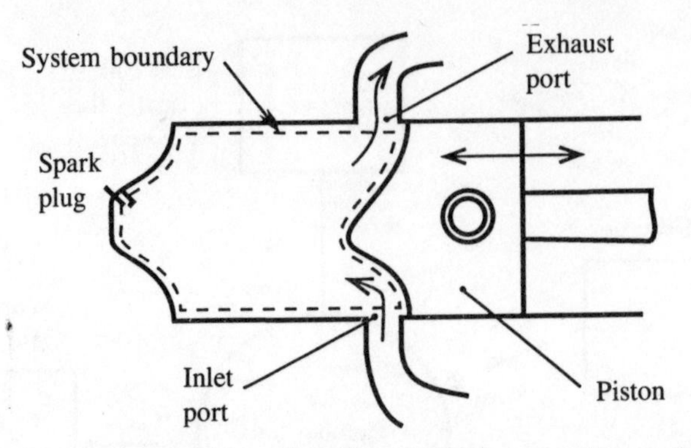

Figure 12.5 *A two-stroke, spark ignition, internal combustion engine.*

1→2 Starting with the piston near bottom-dead-centre and the cylinder full of a combustible mixture of air and petrol vapour at close to ambient temperature and pressure, the piston moves to top-dead-centre. This is known as the compression stroke and the system is closed once the inlet and exhaust ports have been covered by the piston. Both the pressure and the temperature of the air–fuel mixture increase. Work is done on the system. The energy for this comes from the flywheel of the engine.

2→3 The combustible mixture can be regarded as in unstable equilibrium.[13] This means that, although it contains stored chemical energy that could cause a major change in its state, it remains in equilibrium until a suitable stimulus (in this case a spark) is provided to release the chemical energy. Just before the top-dead-centre position a spark occurs at the spark plug and ignites the fuel. Combustion occurs rapidly while the volume of the system changes very little. The air–fuel mixture changes to a mixture of combustion products (mainly H_2O, CO_2 and N_2). The temperature and the pressure within the cylinder rise to their maximum instantaneous values owing to the release of the chemical energy of the fuel.

3→4 The piston moves from top-dead-centre to bottom-dead-centre. Work is done by the system as its volume increases and the pressure and temperature reduce.

4→1 Just before the exhaust port is uncovered by the piston, both the pressure within the cylinder and the temperature are much higher than those of the original air–fuel mixture. When the port is opened the hot gases are discharged to the surroundings. A fresh charge of air and fuel has to be pushed or blown into the cylinder to replace the remaining spent charge. All this occurs while the piston is near the bottom-dead-centre position.

By fitting a pressure sensor in the cylinder head and an angular displacement sensor to the crankshaft, it is possible to measure the variation of the cylinder pressure

13. Unstable equilibrium is defined formally in Chapter 14, section 14.9.1.

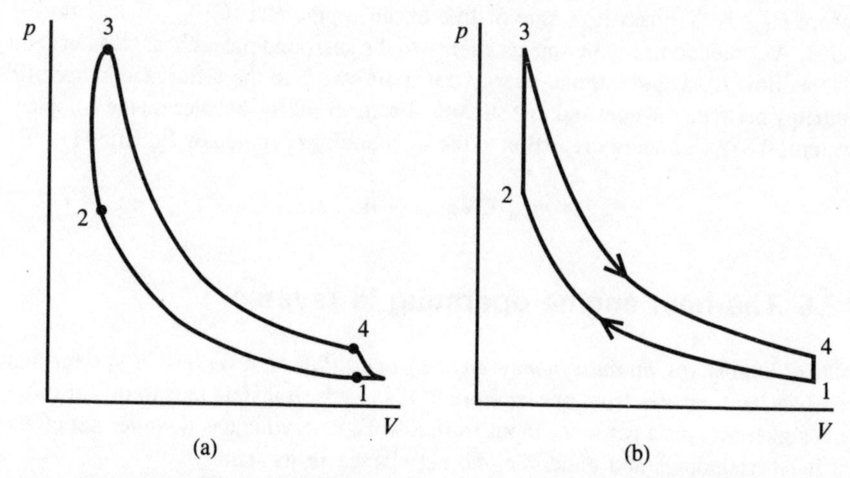

Figure 12.6 *Schematic representations of (a) an experimentally measured p–V diagram for a two-stroke, spark ignition, internal combustion engine, and (b) a p–V equilibrium diagram for an Otto cycle undergone by a closed system that contains only air.*

with the crank position. The crank position, in turn, is related to the instantaneous volume within the cylinder. Hence a p–V diagram (indicator diagram), as shown in Figure 12.6(a), can be plotted from measurements.

In Chapter 17, a true heat engine cycle that is closely related to the actual processes of a spark ignition engine is described. This is shown in Figure 12.6(b). It is known as the air standard Otto cycle. The combustion process $2 \rightarrow 3$ of the actual engine is replaced in the idealized cycle by heat transfer to the system at constant volume. The gas exchange process $4 \rightarrow 1$ is replaced by heat transfer from the closed system at constant volume.

12.4.2 The thermal efficiency of a combustion engine

The *calorific value of a fuel* is the amount of energy released as heat transfer per unit mass of fuel when it is burned in air and the combustion products are allowed to cool to the original temperature of the air and fuel. Values are usually presented for a standard temperature of 25 °C, and for initial and final states at atmospheric pressure. The calorific value at constant pressure can also be described as the decrease in enthalpy when air and the fuel at the reference temperature are allowed to react and the products are brought back to the original temperature and pressure.

The *thermal efficiency of a combustion engine* (irrespective of whether combustion is internal or external) is defined by Equation (12.6):

$$\eta_{\text{th}} = \frac{\dot{W}_{\text{net,out}}}{\dot{m}_{\text{fuel}}(\text{CV})_{\text{fuel}}} \tag{12.6}$$

where $\dot{m}_{fuel}$ is the mass flow rate of fuel to the engine and $(CV)_{fuel}$ is its calorific value. A combustion engine rejects energy to the surroundings as heat transfer, and also as flow work and internal energy transport owing to the difference in specific enthalpy between the inlet and exit streams. From an energy balance on the complete system, the rate of energy rejection to the surroundings is given by Equation (12.7):

$$\dot{E}_{out} = \dot{m}_{fuel}(CV)_{fuel} - \dot{W}_{net,out}. \tag{12.7}$$

12.5 The heat engine operating in reverse

A *heat engine* (or *thermodynamic engine*) *operating in reverse* is a system that receives heat transfer from one or more heat sources, transfers heat to one or more heat sinks, accepts a net work input from a work reservoir, has no other net effect on its surroundings and undergoes no net change in its state.

Normally, the purpose of a heat engine operating in reverse is to transfer heat from a heat source to a heat sink at a higher temperature. The standard representation of a heat engine operating in reverse is shown in Figure 12.7(a). Figure 12.7(b) shows a heat engine operating in reverse where the heat source and sink are both thermal reservoirs. The net work input can be expressed as

$$W_{net,in} = Q_{out} - Q_{in}. \tag{12.8}$$

A *heat pump* is a heat engine operating in reverse whose purpose is to provide a heating effect, i.e. to cause heat transfer Q_{out} to a heat sink. The heating effect is usually provided at temperature levels above the temperature of the environment and the heat source is usually the environment itself, e.g. atmospheric air or ground water.

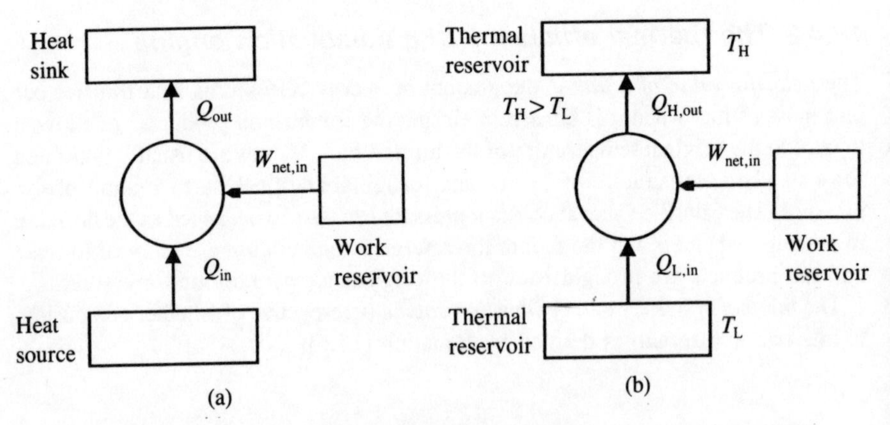

Figure 12.7 *(a) A heat engine operating in reverse, and (b) a heat engine operating in reverse where the heat source and sink are thermal reservoirs.*

A *refrigerator* is a heat engine operating in reverse whose purpose is to provide a cooling effect, i.e. to cause heat transfer Q_{in} from a heat source. The cooling effect is usually provided at temperature levels below the temperature of the environment and the heat sink is usually the environment itself, e.g. atmospheric air or water from a river.

12.5.1 Coefficient of performance, c.o.p.

The ratio of the useful effect (whether heating or cooling) to the required input is used as a convenient figure of merit to describe the performance of heat pumps or refrigerators. This cannot be called an efficiency because its value is usually greater than one for a heat pump and is often greater than one for a refrigerator. For a heat pump

$$\text{c.o.p.}_{hp} = \frac{Q_{out,use}}{W_{net,in}}. \tag{12.9}$$

The subscript 'use' here stands for 'useful' and is to emphasize that in practice only useful heat transfer should be included in a quoted c.o.p. value. In an actual heat pump there may be unwanted heat rejection, e.g. from warm pipework, and in a refrigerator there may be unwanted heat 'pick-up', e.g. to cool pipework. For a refrigerator

$$\text{c.o.p.}_{refr} = \frac{Q_{in,use}}{W_{net,in}}. \tag{12.10}$$

EXAMPLE 12.2
Case study: a domestic refrigerator
Figure 12.8 is a schematic representation of a domestic refrigerator. It consumes 98 W of electric power on average. The average rate of heat transfer to the refrigerant circuit inside the cabinet is 163 W. Calculate the coefficient of performance of the fridge and the rate of heat rejection.

SOLUTION

$$\text{c.o.p.}_{refr} = \frac{Q_{L,in}}{W_{net,in}}$$

Answer

$$= \frac{163\ \text{W}}{98\ \text{W}} = 1.66.$$

$$\dot{Q}_{out} = \dot{Q}_{in} + \dot{W}_{net,in}$$

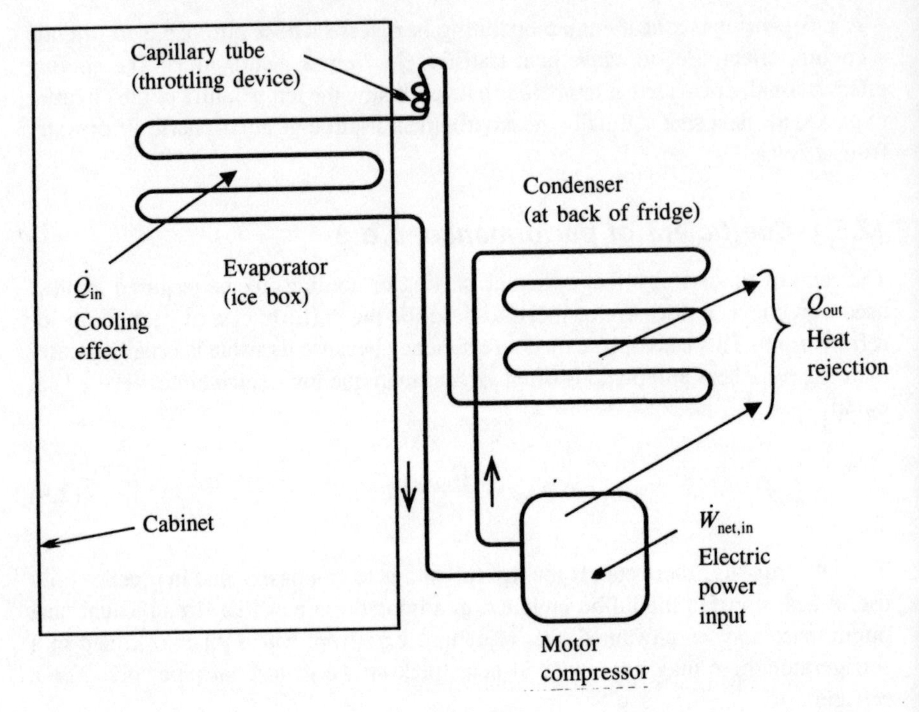

Figure 12.8

Answer

$$= 163 \text{ W} + 98 \text{ W} = 261 \text{ W}.$$

12.6 Summary

Some definitions relating to heat engines and the definition of a heat engine itself have been presented. The heat engine has been described and expressions given for its thermal efficiency. The operation of the two-stroke, spark ignition, internal combustion engine has been outlined and an expression given for the thermal efficiency of combustion engines in general. Heat engines operating in reverse have been described and expressions have been given for their coefficients of performance.

The student should be able to

- *define*
  - — a thermal reservoir, a work reservoir, an equivalent to mechanical shaft work, a mechanical reservoir, a heat engine or thermodynamic engine, the thermal efficiency of a heat engine, a combustion engine, an internal combustion engine, top-dead-centre position, bottom-dead-centre position, the calorific value of a fuel, the thermal efficiency of

a combustion engine, a heat engine or thermodynamic engine operating in reverse, a heat pump, a refrigerator, the c.o.p. of a heat pump, the c.o.p. of a refrigerator

- *sketch*
  — the standard representation of a heat engine, a two-stroke internal combustion engine, an experimentally measured and an ideal indicator diagram for a two-stroke internal combustion engine, the standard representation of a heat engine operating in reverse
- *explain*
  — the distinction between a heat source or sink and a thermal reservoir, the distinction between a work source or sink and a work reservoir, the distinction between a work reservoir and a mechanical reservoir
- *analyze*
  — a system or a plant to determine if it can be considered a heat engine
- *describe*
  — the processes of a two-stroke, spark ignition, internal combustion engine
- *write*
  — the expression for the thermal efficiency of a combustion engine
- *calculate*
  — the thermal efficiency of a heat engine, the c.o.p. of a refrigerator or heat pump, the thermal efficiency of a combustion engine.

12.7 Self-assessment questions

12.1 The rate of heat transfer to a heat engine is 49 kW and its net power output is 9.4 kW. Determine the thermal efficiency and the rate of heat rejection to the surroundings.

12.2 Which items in the following list fully meet the definition of a heat engine or a heat engine operating in reverse? If possible, select a system boundary in each case to enclose a system that could be considered either a heat engine or a heat engine operating in reverse. In the case of a heat engine operating in reverse, state whether it is a heat pump or a refrigerator.

(a) A two-stroke diesel engine. In this engine, air is taken into the cylinder and compressed during the compression stroke of the reciprocating piston. The fuel oil is injected into the compressed air at high pressure and combustion occurs spontaneously. After expan-

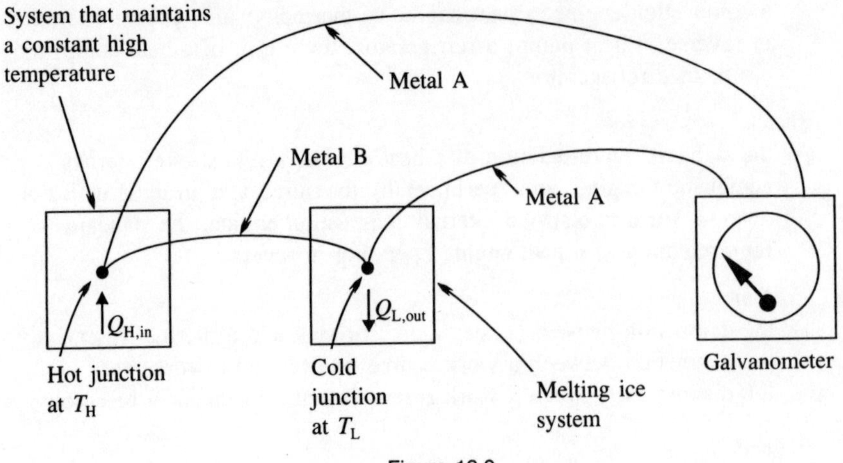

Figure 12.9

sion in the expansion stroke, the combustion products, still at quite a high temperature, are discharged to the atmosphere.

(b) A thermocouple connected to a galvanometer, as shown schematically in Figure 12.9. The arrangement consists of a hot junction in a system that maintains a constant high temperature, a cold junction immersed in melting ice and a moving-coil type of galvanometer, which detects the small current produced through its own small resistance whenever the hot junction is at a temperature that differs from the cold junction temperature.

(c) The disc brake of a motorcycle while it provides a constant braking force as the motorcycle descends a hill at constant speed. This absorbs work and transfers heat to the surroundings.

(d) A steam locomotive. Water is evaporated in a boiler to produce steam. The steam is expanded to atmospheric pressure in a cylinder that has a reciprocating piston. At the end of the expansion process the steam is discharged to the atmosphere.

(e) A wind-driven churn for heating water that enters and leaves the churn in steady flow.

(f) A closed, steady flow, vapour compression plant consisting of an evaporator, a motor-driven compressor, a condenser and a throttle-type expansion device. This provides cooling of an ice rink and heating for a swimming pool.

12.3 Calculate the required power input to a refrigerator if it operates with a c.o.p. of 1.75 and provides a cooling effect of 41 kW. Determine the rate of heat transfer to the surroundings.

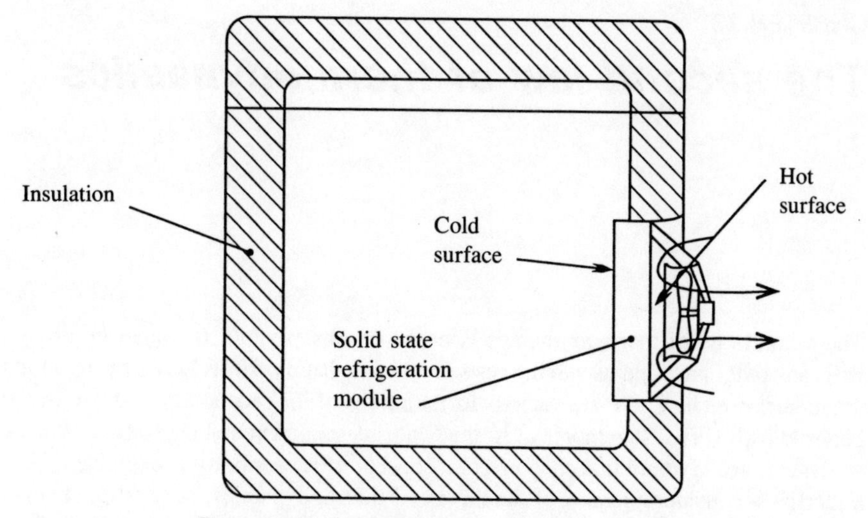

Figure 12.10 *An electronically refrigerated cold box.*

12.4 An electronically refrigerated coolbox for use in a car is illustrated in Figure 12.10. It consumes a total of 28.1 watts of d.c. electric power from the cigar-lighter socket. Of this amount the fan consumes 9.5 watts. The cooling effect, which is the rate of heat transfer from the contents of the box to the internal cold surface, is 28.5 watts. What is the c.o.p. of the device? This should be based on the useful cooling effect and the *total* power input to the device. What is the rate of heat rejection from the hot surface of the solid state refrigeration module?

12.5 Calculate the thermal efficiency of an internal combustion engine that burns fuel at the rate of 4.11 kg/h while it produces 11.43 kW of output power. The calorific value of the fuel is 42 500 kJ/kg. Calculate also the rate of energy transfer to the surroundings from the engine.

CHAPTER 13

The second law of thermodynamics

The second law of thermodynamics is based on observation. It cannot be proved from any other laws and its validity rests on the fact that no effects have been observed that contravene it. There are various formulations of the second law and two of the better known verbal statements of it, the Clausius statement and the Kelvin–Planck statement, are given in this chapter. The concept of reversibility is explained. This underlies a considerable part of the theory of thermodynamics. A principle known as Carnot's principle is stated and proved from one of the second-law statements – this is an important introduction to the reasoning that underlies the application of the second law of thermodynamics. The thermodynamic temperature scale, which is founded on the second law of thermodynamics, is defined. The performance of a reversible heat engine is described, as is the performance of a reversible heat engine operating in reverse. The Clausius inequality, which is a mathematical statement of the second law, is stated and derived. It is shown from this how the limiting thermal efficiency of any heat engine can be determined.

The implications of the second law are very far reaching. For instance, not all processes that would satisfy the first law of thermodynamics can occur. The second law of thermodynamics and deductions from it, such as Carnot's principle, provide the further necessary criteria that must be satisfied by processes that can occur.

13.1 The Clausius statement

It is impossible for a system to cause heat transfer from a thermal reservoir to another thermal reservoir at a higher temperature, to have no other net effect on its surroundings and to undergo no net change.

Figure 13.1(a) illustrates the type of situation that is not permitted according to the Clausius statement of the second law. The circle in this diagram represents a system that undergoes no net change when heat transfer occurs to the system from a thermal reservoir and from the system to a thermal reservoir at a higher temperature. Figure 13.1(b) shows a type of situation that differs from that in (a) only in that the directions of the heat transfer interactions are reversed. The Clausius statement of the second law does not exclude it: heat transfer through the walls of a building from warm air inside to cold air outside is an example. Figure 13.1(c) shows a situation where heat transfer does occur from a thermal reservoir and to a thermal

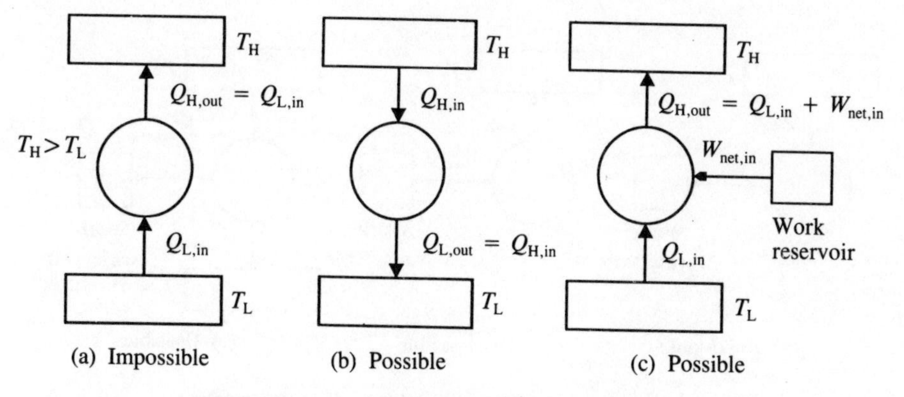

Figure 13.1 *Illustration of the Clausius statement.*

reservoir at a higher temperature. This situation differs from that shown in (a) in so far as there is a net input of work to the system from a work reservoir and the heat output is equal to the sum of the work input and the heat input. Systems of this type can be built and are not in conflict with the Clausius statement: a domestic refrigerator is a common example.

13.2 The Kelvin–Planck statement

It is impossible for a system to accept heat transfer from a thermal reservoir and provide a net work output to a work reservoir, to have no other net effect on its surroundings and to undergo no net change.

Figure 13.2(a) shows the type of situation that is excluded by the Kelvin–Planck statement of the second law. It shows a device that has no net effect on the surroundings other than the conversion of heat transfer to work while it undergoes no net change. The situation shown in Figure 13.2(b) is similar to that in (a) but the directions of the energy interactions are reversed. The system has no effect on the surroundings other than the conversion of work to heat transfer while it undergoes no net change. This situation commonly exists and is not excluded by the Kelvin–Planck statement of the second law. A brake on a bicycle is an example: the work done by the bicycle wheel becomes heat transfer to the surroundings. The situation shown in Figure 13.2(c) involves heat transfer from a thermal reservoir to a system that undergoes no net change and that provides a net work output to a work reservoir. However, it differs from the situation in (a) because there is also heat rejection to another thermal reservoir at a lower temperature. As energy is conserved, the net work output is less than the heat input. Systems of this type can be devised and are not in conflict with the Kelvin–Planck statement: a steam power plant for generating electricity is a common example.

It can be shown that if either the Clausius statement or the Kelvin–Planck statement

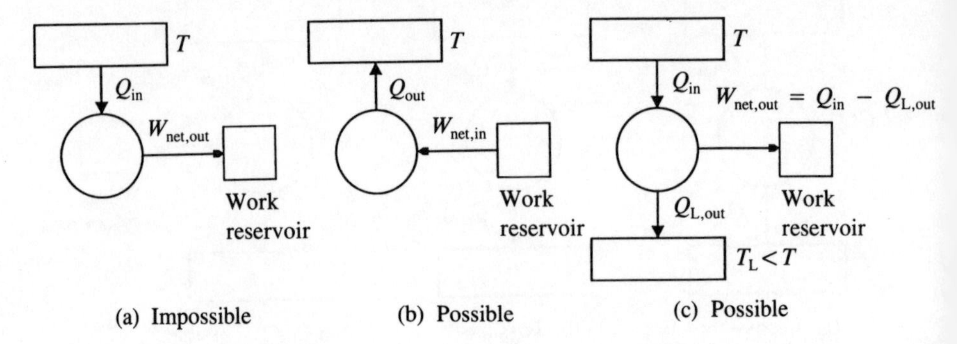

Figure 13.2 *Illustration of the Kelvin–Planck statement.*

of the second law is true, then the other statement is also true. This is left as an exercise for the reader. The two statements are thus equivalent.

13.3 Immediate implications of the second law

13.3.1 The directionality of heat transfer

From the Clausius statement, heat transfer can only occur spontaneously in the direction of temperature decrease. If two systems at different temperatures are brought together as a composite system and enclosed within an isolating boundary, heat transfer will occur from the hotter subsystem to the cooler subsystem. The state of the system will change as long as any temperature differences remain. This non-equilibrium process cannot reverse at any point, as this would require heat transfer in the direction of temperature increase. It can only proceed until all temperature differences are eliminated: the state of thermal equilibrium.

13.3.2 The non-equivalence of heat and work

In terms of the first law, heat and work are equivalent in that they are both forms of energy transfer. However, from the Kelvin–Planck statement of the second law, heat and work are fundamentally different modes of energy transfer, or effects. While a system that converts work to an equivalent energy transfer as heat is possible, a device that converts heat to an equivalent energy transfer as work is impossible.

13.4 Reversibility

The words 'reversible' and 'reversibility' have a special and very precise meaning in thermodynamics. Reversible processes would exist in a thermodynamically ideal world. They are important because they allow standards to be set for ideal performance, against which the performance of real devices can be judged.

A *process* of a system is *reversible* if at any point the system could be returned to its initial state in such a way that there would be no net effect on the surroundings. An *irreversible process* is a process that is not reversible.[14]

13.4.1 Points to note about reversible processes

1 There must be no viscous forces or turbulence within the system, e.g. any process involving 'stirring work' would be irreversible.
2 There must be no friction work interactions between subsystems of the system.
3 If there is heat transfer between any two subsystems of the system, there can be only infinitesimal temperature differences between those subsystems.
4 A reversible process is a limiting case, which can be approached but not realized.
5 If a system can be brought back to its initial state after a process has occurred, this is not a sufficient condition for the process to be described as reversible. The necessary further condition is that when the process has been reversed any effects on the surroundings are nullified.

It is deduced in Chapter 14, section 14.9.1, that an equilibrium process is a special case of a reversible process. Therefore, processes that have been described as equilibrium processes up to this point are also reversible processes.

13.5 Carnot's principle

Carnot's principle consists of two parts, as follows:

1 *The thermal efficiencies of all reversible heat engines that operate between the same two thermal reservoirs are the same.*
2 *The thermal efficiency of a reversible heat engine is greater than the thermal efficiency of an irreversible heat engine when both engines operate between the same two thermal reservoirs.*

13.6 Proof of Carnot's principle

To prove statements 1 and 2, assume they are false. Let A and B in Figure 13.3 be two heat engines. Let A, at least, be reversible and let the thermal efficiency of B be higher than that of A. Let both engines produce the same net work output.

14. The reader might find it helpful to re-read the description of the candle-lighting process in section 1.1 of Chapter 1.

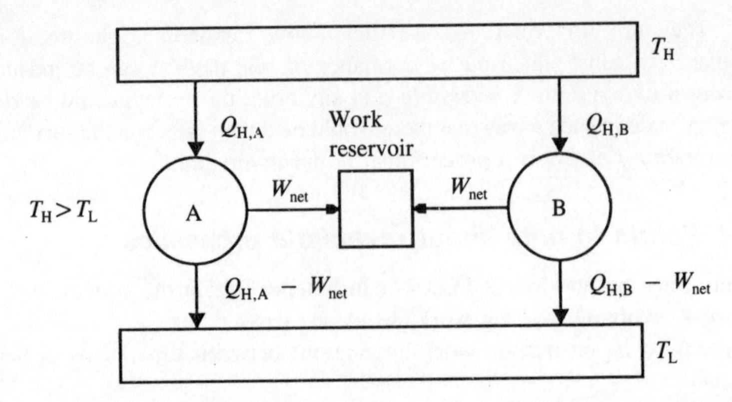

Figure 13.3 *Two engines,* A *and* B*, that operate between the same thermal reservoirs and produce the same net work output.*

$$E_{th,B} > E_{th,A},$$

i.e.

$$\frac{W_{net}}{Q_{H,B}} > \frac{W_{net}}{Q_{H,A}}$$

$$\therefore\ Q_{H,B} < Q_{H,A}.$$

As A is reversible, its work and heat transfer interactions can be reversed as in Figure 13.4.

This is equivalent to the device shown in Figure 13.5, where $Q_{H,A} > Q_{H,B}$. But this violates the Clausius statement. Therefore, the assumption that heat engine B has the higher efficiency cannot be valid and the following conclusions can be drawn:

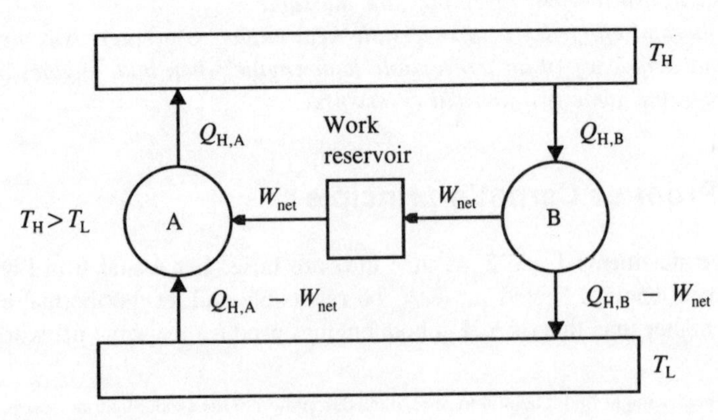

Figure 13.4

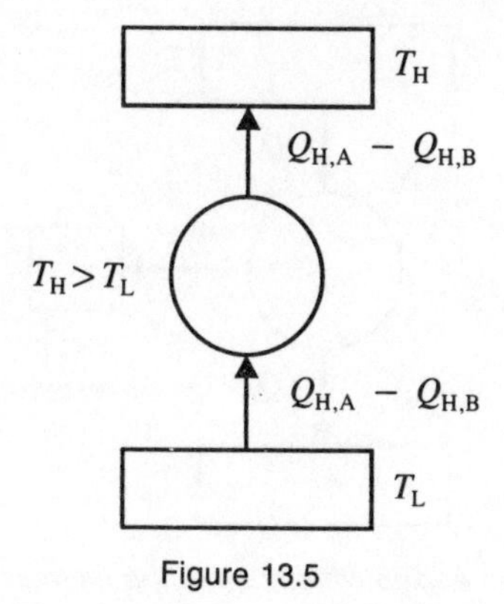

Figure 13.5

1 If both heat engines are reversible $\eta_{th,B} \le \eta_{th,A}$, but, by switching A and B and by the same arguments, $\eta_{th,A} \le \eta_{th,B}$. Therefore, the two reversible heat engines must have the same thermal efficiency and statement 1 of Carnot's principle is true.
2 If heat engine B is not reversible it can be concluded that $\eta_{th,B} \le \eta_{th,A}$. However, if the thermal efficiencies are equal, heat engine B is reversible (because its net effects on its surroundings can be reversed by engine A operating in reverse). Therefore, when B is not reversible, $\eta_{th,B} < \eta_{th,A}$, and so statement 2 of Carnot's principle is true.

13.7 The thermodynamic or absolute temperature scale

The fact that all reversible heat engines operating between the same two thermal reservoirs have the same thermal efficiency allows the construction of a temperature scale based on the characteristics of a reversible heat engine and on one fixed reference temperature: the triple point temperature of water. This scale is known as the *thermodynamic or absolute temperature scale*.

Figure 13.6 shows a reversible heat engine where one of the thermal reservoirs is at the triple point temperature of water. This will operate in the normal direction if $T > T_{tp}$ and in reverse if $T < T_{tp}$. Whether it operates in the normal or reverse direction, the following relationship, which is based on the energy balance, applies:

$$\eta_{rev} = \frac{W_{net}}{Q} = \frac{Q - Q_{tp}}{Q} = 1 - \frac{Q_{tp}}{Q}. \qquad (13.1)$$

In Equation (13.1), Q and Q_{tp} will have positive values if the reversible engine operates in the normal direction and negative values if it operates in reverse. Carnot's principle states that all reversible heat engines that operate between the same thermal reservoirs have the same thermal efficiency. For any combination of the temperatures

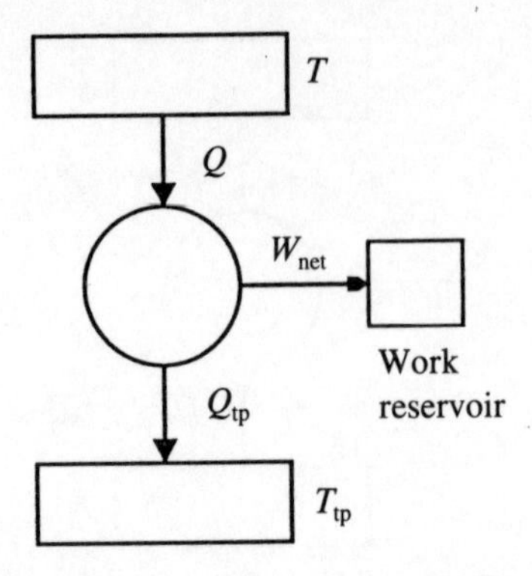

Figure 13.6 *A reversible heat engine where one of the thermal reservoirs is at the triple point temperature.*

of the two thermal reservoirs, the thermal efficiency has a unique value. Therefore,

$$1 - \frac{Q_{tp}}{Q} = \eta_{th,rev} = f(T, T_{tp}) \tag{13.2}$$

where f is a function of T and T_{tp}.

In general, if a variable x is a function of two variables y and z, then any of the three variables can be expressed as a function of the other two. In addition to the original function, two other functions can be written in which one of the variables is expressed explicitly in terms of the other two. This is stated in equation form as follows:

if $$x = f(y, z)$$

then $$y = f'(x, z)$$

and $$z = f''(x, y)$$

where f′ and f″ are functions that can be determined from function f.

Therefore, from Equation (13.2), the absolute temperature T can be expressed as a function of the ratio Q_{tp}/Q and the triple point temperature T_{tp}:

$$T = f'\left(\frac{Q_{tp}}{Q}, T_{tp}\right) \tag{13.3}$$

where f′ is a function of Q_{tp}/Q and T_{tp}.

The definition of a temperature scale based on Equation (13.3) involves specifying the form of the function f′ and the value to be assigned to the triple point temperature. Any number of scales could be defined by specifying different functions and different values for the triple point temperature. The absolute temperature scale has been defined by specifying this function as

$$T = f'\left(\frac{Q_{tp}}{Q}, T_{tp}\right) \triangleq T_{tp}\frac{Q}{Q_{tp}}. \tag{13.4}$$

Thus

$$\frac{T}{T_{tp}} = \frac{Q}{Q_{tp}}. \tag{13.5}$$

The value 273.16 K has been assigned to the triple point temperature of water. This results in the temperature difference between the boiling and freezing points of water at standard atmospheric pressure being 100 K. It thus brings the absolute temperature scale, based on the reversible engine and the triple point temperature of water, into line with the absolute scale and the conventional scale (the Celsius scale) that had previously been based on the constant volume gas thermometer and the boiling and freezing points of water at standard atmospheric pressure. By international agreement the absolute scale based on the reversible engine is now the definitive absolute scale. In the SI system its unit is the kelvin. The conventional Celsius scale is now defined in terms of the absolute kelvin scale (Chapter 4, section 4.4).

Equation (13.4) allows thermometers to be calibrated very precisely with respect to the absolute temperature scale. The heat transfer quantities in the equation are experimentally determined from measurable properties of substances and not by the implementation of a reversible heat engine: this would be impossible as all real processes are irreversible. The procedures for the calibration of reference thermometers are highly specialized and are beyond the scope of this book.

13.8 The thermal efficiency or c.o.p. of a reversible heat engine

A reversible heat engine can operate in the normal direction as a heat engine, or in reverse as a heat pump or refrigerator. It is valid, therefore, to describe the thermal efficiency, heat pump c.o.p. and refrigeration c.o.p. for a reversible engine. Real engines, heat pumps or refrigerators usually cannot operate in both directions.[15] Figure 13.7 shows two reversible heat engines, H and L, which together provide a net work output to a work reservoir. One operates between thermal reservoirs

15. Certain Stirling cycle machines can function as an engine to provide power, or as an engine operating in reverse to provide cooling or heating. The thermal efficiency or c.o.p. is always lower than that of a reversible engine. There is an important distinction between an engine that can operate in reverse and a (*thermodynamically*) reversible engine.

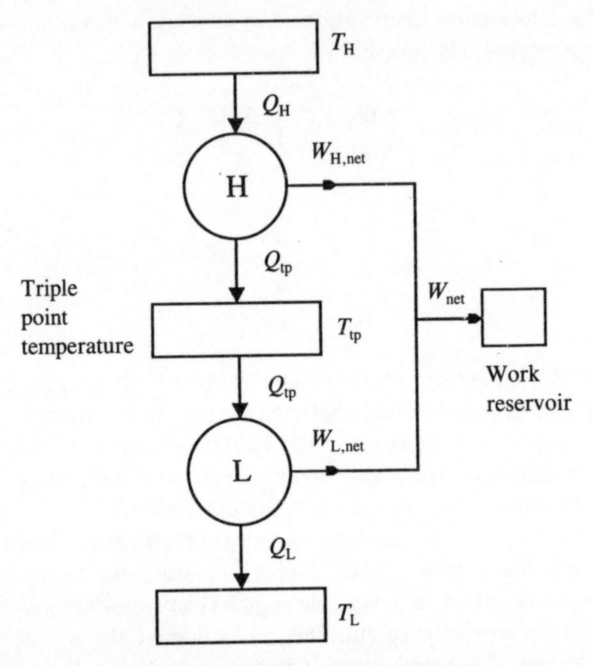

Figure 13.7 *Two reversible heat engines with a common thermal reservoir at the triple point temperature. The net heat transfer to this common thermal reservoir is zero ($Q_{tp} - Q_{tp}$). Together these two engines provide net work to a work reservoir.*

at T_H and T_{tp}. The other operates between thermal reservoirs at T_{tp} and T_L. Subject to the constraint that $T_H > T_L$, both T_H and T_L can be higher than or lower than T_{tp}. If T_H is less than T_{tp}, then the heat and work quantities associated with heat engine H (defined by the arrows in Figure 13.7) would have negative values and the heat engine would be operating in reverse. Similarly, reversible heat engine L would be operating in reverse if T_L is greater than T_{tp}.

Any reversible heat engine that operates between thermal reservoirs at temperatures T_H and T_L must have the same thermal efficiency as the combination of two reversible heat engines shown in Figure 13.7, as together the two heat engines, H and L, represent a reversible heat engine that operates between the thermal reservoirs at T_H and T_L. There is no net heat transfer to the thermal reservoir at the triple point temperature and so it could be eliminated.

From the definition of the absolute temperature scale, Equations (13.6) and (13.7) apply for reversible heat engines H and L respectively, irrespective of whether each is operating in the normal or the reverse direction.

$$T_H = T_{tp}\frac{Q_H}{Q_{tp}} \quad \therefore \; Q_H = Q_{tp}\frac{T_H}{T_{tp}} \tag{13.6}$$

$$T_L = T_{tp}\frac{Q_L}{Q_{tp}} \quad \therefore \; Q_L = Q_{tp}\frac{T_L}{T_{tp}}. \tag{13.7}$$

Hence

$$\frac{Q_L}{Q_H} = \frac{T_L}{T_H}. \tag{13.8}$$

The thermal efficiency of a reversible heat engine operating between two thermal reservoirs at temperatures T_H and T_L, where $T_H > T_L$, is therefore given by

$$\eta_{th,rev} = \frac{Q_H - Q_L}{Q_H} = \frac{T_H - T_L}{T_H}. \tag{13.9}$$

For a reversible heat engine operating in reverse:

$$c.o.p._{rev,hp} = \frac{Q_H}{Q_H - Q_L} = \frac{T_H}{T_H - T_L}, \tag{13.10}$$

$$c.o.p._{rev,refr} = \frac{Q_L}{Q_H - Q_L} = \frac{T_L}{T_H - T_L}. \tag{13.11}$$

The latter three expressions are important limiting cases for heat engines, heat pumps and refrigerators that operate between two thermal reservoirs. The performance of a reversible heat engine can never be achieved in practice. All real engines, heat pumps or refrigerators involve irreversible processes such as heat transfer with finite temperature differences, fluid friction or friction work interactions between subsystems.

EXAMPLE 13.1

What is the maximum theoretical c.o.p. of a refrigeration plant that maintains a storage vessel at −174.5 °C when the ambient temperature is 21.7 °C?

SOLUTION

$$T_H = 21.7\,[°C] + 273.15\,[K] = 294.85\ K$$

$$T_L = -174.5\,[°C] + 273.15\,[K] = 98.65\ K$$

$$c.o.p._{rev,refr} = \frac{T_L}{T_H - T_L} = \frac{98.65\,[K]}{(294.85 - 98.65)\,[K]}$$

Answer

$$= 0.503.$$

13.9 The inequality of Clausius

The inequality of Clausius, which is a mathematical statement of the second law, is presented as a precursor to the definition of a difference in a property known as entropy. It is derived here from Equation (13.8), which follows from the definition of the absolute temperature scale, and the Kelvin–Planck statement of the second law.

Figure 13.8 shows a system S that undergoes no net change, provides a net work output to a work reservoir, has heat transfer interactions with n thermal reservoirs that have different temperatures and has no other net effect on its surroundings. The heat transfer arrows for this system are shown in the conventional positive direction. As there is a net work output, the net heat transfer for the system is positive, but some of the heat transfer interactions may be negative.

Figure 13.9 shows the same system with the same work and heat interactions. The n thermal reservoirs are linked to a reference thermal reservoir at temperature T_r by n reversible heat engines operating in the normal direction or the reverse direction, as appropriate, so that the net heat transfer to each of the n thermal reservoirs from system S and the interacting reversible engine is zero. The heat transfer and work arrows of the reversible engines indicate the assumed positive directions for their interactions: some will have positive values and some will be negative.

From an energy balance for system S

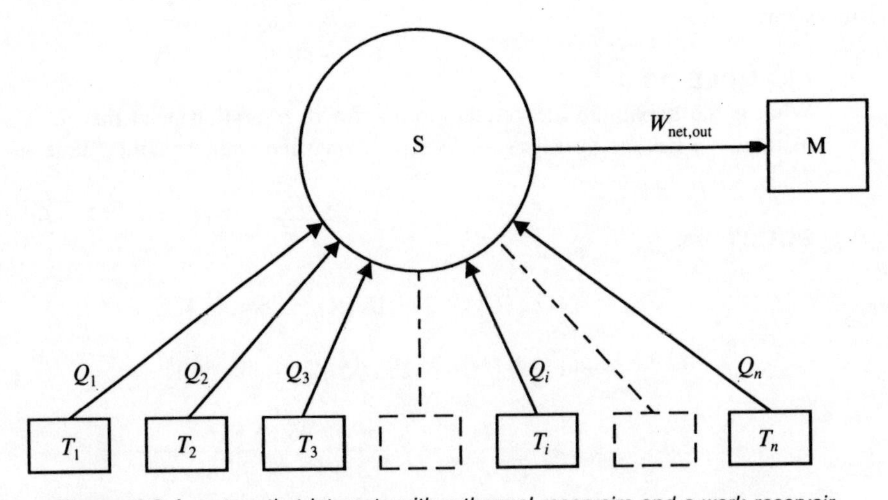

Figure 13.8 *A system that interacts with n thermal reservoirs and a work reservoir.*

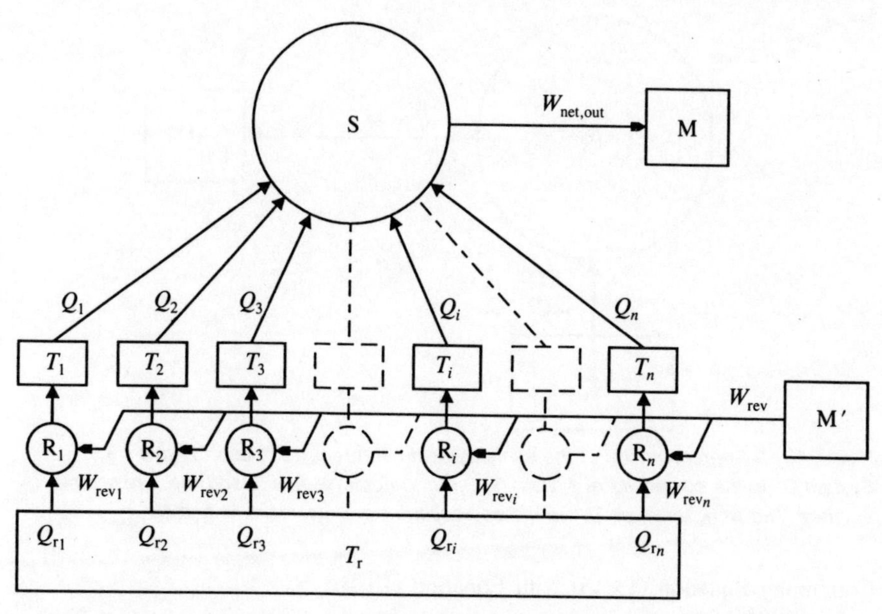

Figure 13.9 *The same system and reservoirs as in Figure 13.8 but with the thermal reservoirs linked to a reference thermal reservoir by reversible heat engines that interact with another work reservoir* M′.

$$W_{net,out} = \sum_{i=1}^{n} Q_i. \tag{13.12}$$

For a reversible heat engine R_i that interacts with one of the n thermal reservoirs

$$\frac{Q_i}{Q_{r_i}} = \frac{T_i}{T_r} \tag{13.13}$$

$$Q_{r_i} = Q_i \frac{T_r}{T_i}. \tag{13.14}$$

Hence

$$\sum_{i=1}^{n} Q_{r_i} = \sum_{i=1}^{n} Q_i \frac{T_r}{T_i} = T_r \sum_{i=1}^{n} \frac{Q_i}{T_i}. \tag{13.15}$$

From an energy balance for the n reversible engines

$$W_{rev} = \sum_{i=1}^{n} Q_i - \sum_{i=1}^{n} Q_{r_i}. \tag{13.16}$$

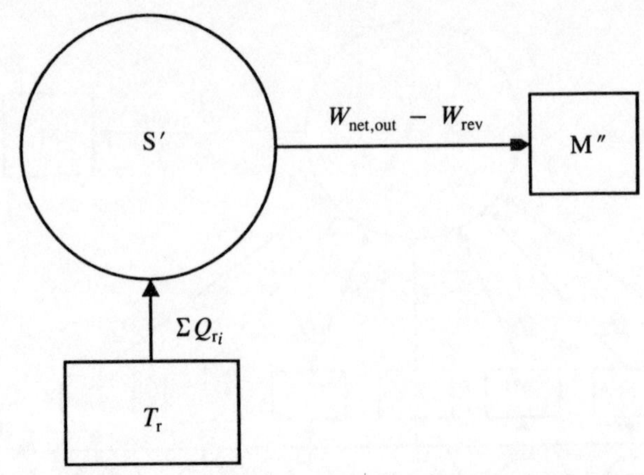

Figure 13.10 *Representation of the same systems and reservoirs as in Figure 13.9. System* S′ *is the composite of system* S, *the n thermal reservoirs and the n reversible engines. The work reservoir* M″ *is the composite of the reservoirs* M *and* M′.

Combining Equation (13.15) with Equation (13.16)

$$W_{\text{rev}} = \sum_{i=1}^{n} Q_i - T_{\text{r}} \sum_{i=1}^{n} \frac{Q_i}{T_i}. \tag{13.17}$$

The composite of system S, the n thermal reservoirs and the n reversible engines is a system that undergoes no net change, interacts with a thermal reservoir at temperature T_{r} and with a work reservoir, which is the composite of the original two work reservoirs M and M′, and has no other net effect on its surroundings. This is shown in Figure 13.10.

As the composite system S′ must satisfy the Kelvin–Planck statement of the second law, its net work output cannot be positive. Therefore

$$W_{\text{net,out}} - W_{\text{rev}} \leq 0. \tag{13.18}$$

Using Equation (13.12) to substitute for $W_{\text{net,out}}$ and Equation (13.17) to substitute for W_{rev},

$$\sum_{i=1}^{n} Q_i - \left(\sum_{i=1}^{n} Q_i - T_{\text{r}} \sum_{i=1}^{n} \frac{Q_i}{T_i} \right) \leq 0. \tag{13.19}$$

Hence

$$T_{\text{r}} \sum_{i=1}^{n} \frac{Q_i}{T_i} \leq 0. \tag{13.20}$$

And, as T_r is positive,

$$\sum_{i=1}^{n} \frac{Q_i}{T_i} \leq 0. \tag{13.21}$$

Equation (13.18), and hence Equation (13.21), applies as an equality only if $W_{net,out} = W_{rev}$. If this is so, then, from Equation (13.12), $W_{rev} = \Sigma_{i=1}^{n} Q_i$ and then, from Equation (13.16), $\Sigma_{i=1}^{n} Q_{r_i} = 0$. Hence, the reference thermal reservoir at temperature T_r can be eliminated. The composite system of the n reversible engines interacts with the same set of thermal reservoirs as system S. This composite system of reversible engines can nullify the net effects of system S on the thermal reservoirs and on the original work reservoir M. Therefore, system S undergoes, or contains, only reversible processes.

Equation (13.21) applies for any number of thermal reservoirs between one and infinity. The positive direction for heat transfer in the equation is the conventional positive direction: into the system S. The interactions with the thermal reservoirs can occur simultaneously, as in a steady flow cycle, or sequentially in time, as in a cycle of a closed system that contains a compressible substance. In general, the amounts of heat transfer and the temperature levels at which they occur can vary over the boundary surface and with time. In practical situations the temperature at the boundary of a system usually varies continuously rather than discretely with time and over the surface area. This is equivalent to an infinite number of thermal reservoirs as there are an infinite number of temperatures between any two values. To evaluate the summation in the most general case, the surface of the boundary must be broken down into a sufficiently large number of small subdivisions so that temperature variations over any subdivision are negligible. Also, the time interval during which no net change of the system occurs must be broken down into a sufficiently large number of small steps so that the temperature change at any point on the surface is negligible over one time step. The summation can be evaluated if the heat transfer and the temperature for each subdivision of the boundary are known for each time step. If the subdivisions and the time step are made infinitesimal, the summation can be replaced by the mathematical notation for an integral.

The Clausius inequality applies for any system that has heat transfer or work interactions with other systems, has no other net effect on its surroundings and undergoes no net change. It is written as the following two equations:

$$\int \frac{dQ}{T} = 0 \quad \text{for a system that undergoes or contains only reversible processes} \tag{13.22}$$

$$\int \frac{dQ}{T} < 0 \quad \text{for a system that undergoes or contains irreversible processes.} \tag{13.23}$$

The inequality can also be written as a single expression:

$$\int \frac{dQ}{T} \leq 0. \tag{13.24}$$

A more restricted form of the Clausius inequality is given in Equations (13.25) and (13.26). The circle around the integration sign indicates that it is the integral around a cycle. This form applies for a cycle undergone by a closed system where the incremental amount of heat transfer and the temperature vary with time only. It also applies for a point system that undergoes a cycle as it passes around a steady flow system.

$$\oint \frac{dQ}{T} = 0 \quad \text{for a reversible cycle} \tag{13.25}$$

$$\oint \frac{dQ}{T} < 0 \quad \text{for an irreversible cycle.} \tag{13.26}$$

This form can also be written as the single equation

$$\oint \frac{dQ}{T} \leq 0. \tag{13.27}$$

13.9.1 The impossibility of a thermal efficiency of 100%

From Equation (12.5) of Chapter 12, a thermal efficiency of 100% for a heat engine is only possible if there is no heat rejection whatsoever. A thermal efficiency of 100% is therefore specifically excluded by the second law of thermodynamics, in the form of the Kelvin–Planck statement, where the only heat source is an isothermal reservoir.

In general, however, a heat engine can accept heat transfer at various discrete temperatures or over various temperature ranges. Similarly, it can reject heat at various temperatures or over various temperature ranges. The Clausius inequality can be written in the form of Equation (13.28) or (13.29) where Q_{in} represents positive heat transfer, or heat input, and Q_{out} represents negative heat transfer, or heat rejection:

$$\int \frac{dQ_{in}}{T} - \int \frac{dQ_{out}}{T} \leq 0 \tag{13.28}$$

$$\int \frac{dQ_{in}}{T} \leq \int \frac{dQ_{out}}{T}. \tag{13.29}$$

If the absolute temperature is positive,[16] the left-hand side of Equation (13.28) can only be negative if there is heat rejection. From the Clausius inequality it follows that, even for the most general case, a thermal efficiency of 100% is excluded.

13.9.2 The limiting thermal efficiency of a general heat engine

Let the maximum temperature at which heat transfer occurs to an engine be $T_{H,max}$ and let $T_{L,min}$ be the minimum temperature at which heat transfer occurs from the engine. The following inequalities can be written.

As

$$\frac{\delta Q_{in}}{T_{max}} \leq \frac{\delta Q_{in}}{T} \quad \text{for all } T \leq T_{max} \tag{13.30}$$

$$\int \frac{dQ_{in}}{T_{max}} \leq \int \frac{dQ_{in}}{T} \tag{13.31}$$

and, as T_{max} is fixed,

$$\frac{Q_{in}}{T_{max}} \leq \int \frac{dQ_{in}}{T}. \tag{13.32}$$

Also, as

$$\frac{\delta Q_{out}}{T} \leq \frac{\delta Q_{out}}{T_{min}} \quad \text{for all } T \geq T_{min} \tag{13.33}$$

$$\int \frac{dQ_{out}}{T} \leq \int \frac{dQ_{out}}{T_{min}} \tag{13.34}$$

and

$$\int \frac{dQ_{out}}{T} \leq \frac{Q_{out}}{T_{min}}. \tag{13.35}$$

Hence, from Equations (13.29), (13.32) and (13.35)

$$\frac{Q_{in}}{T_{max}} \leq \int \frac{dQ_{in}}{T} \leq \int \frac{dQ_{out}}{T} \leq \frac{Q_{out}}{T_{min}}. \tag{13.36}$$

16. In normal engineering applications, negative absolute temperatures do not exist. Absolute zero temperature can be approached but not reached.

Hence

$$Q_{out} \geq Q_{in}\frac{T_{min}}{T_{max}} \tag{13.37}$$

$$\eta_{th} = \frac{Q_{in} - Q_{out}}{Q_{in}} \leq \frac{Q_{in} - Q_{in}(T_{min}/T_{max})}{Q_{in}}. \tag{13.38}$$

Therefore,

$$\eta_{th} \leq \frac{T_{max} - T_{min}}{T_{max}}. \tag{13.39}$$

In a similar way, it can be shown that the expression for the c.o.p. of a reversible heat pump or refrigerator, evaluated in terms of the maximum temperature at which heat input occurs (which replaces T_L) and the minimum temperature at which heat output occurs (which replaces T_H), is the limiting value for a general heat pump or refrigerator.

EXAMPLE 13.2

What is the maximum theoretical thermal efficiency of an engine that receives heat transfer at temperatures no higher than 950 °C and rejects heat at temperatures no lower than 35 °C?

SOLUTION

$$T_{max} = 950\,[°C] + 273.15\,[K] = 1223.15\,K$$

$$T_{min} = 35\,[°C] + 273.15\,[K] = 308.15\,K$$

$$\eta_{th,max} = \frac{T_{max} - T_{min}}{T_{max}} = \frac{(1223.15 - 308.15)\,[K]}{1223.15\,[K]}$$

Answer

$$= 0.748 = 74.8\%.$$

13.10 Practical tips

- The temperatures in the expressions for the thermal efficiency of an ideal heat engine and the c.o.p. of an ideal heat engine operating in reverse must be in absolute units.
- Diagrams are very helpful in explaining the reasoning associated with the

second law of thermodynamics. Make sure to use diagrams as part of the reasoning process.

- In diagrams relating to heat engines and heat engines operating in reverse, the directions of the arrows indicate the positive directions for the heat and work interactions. The symbols for the heat and work quantities are shown beside the arrows. These arrows override the sign convention (that energy transfer into a system is positive) for the heat and work quantities they represent.

13.11 Summary

The Clausius statement and the Kelvin–Planck statement of the second law of thermodynamics have been described in words and explained with the aid of schematic diagrams. The concept of reversibility has been explained. Carnot's principle has been stated and proved from the Clausius statement of the second law. Using Carnot's principle, it has been shown that any temperature can be expressed as a function of a reference temperature and of the ratio of the heat transfer quantities of a reversible heat engine operating between the temperature of interest and the reference temperature. Hence, the thermodynamic temperature scale has been defined. Expressions for the thermal efficiency and for the heat pump and refrigeration coefficients of performance of reversible heat engines that operate between two thermal reservoirs have been derived in terms of the absolute temperatures of the reservoirs. The inequality of Clausius has been stated and derived. An expression for the limiting thermal efficiency of a general heat engine has been derived.

The student should be able to

- *state*
  — the Clausius statement of the second law of thermodynamics in words, the Kelvin–Planck statement of the second law of thermodynamics in words, Carnot's principle in two parts

- *define*
  — a reversible process, an irreversible process

- *sketch*
  — diagrams to explain what the Clausius statement permits and prohibits, diagrams to explain what the Kelvin–Planck statement permits and prohibits

- *prove*
  — the two statements of Carnot's principle from the Clausius statement of the second law

- *show*
  — that any given temperature can be expressed as a function of a reference temperature and the ratio of the heat transfer quantities of a

reversible heat engine that operates between that temperature and the reference temperature
— from the Clausius inequality that no general heat engine can have a thermal efficiency of 100%

- *explain*
  — how the absolute temperature scale is defined, some factors that could cause a process to be irreversible

- *derive*
  — the expressions for the thermal efficiency and the c.o.p. of a reversible heat engine that operates between two thermal reservoirs given the defining equation for the absolute temperature scale, the Clausius inequality, an expression for the limiting thermal efficiency of a general heat engine in terms of the maximum and minimum temperatures at which heat input and output occur respectively

- *apply*
  — the expressions for the ideal or limiting thermal efficiency of a heat engine, the expressions for the ideal or limiting c.o.p. of a heat pump or refrigerator.

13.12 Self-assessment questions

13.1 The rate of heat transfer to a reversible heat engine is 45.6 kW. It operates between thermal reservoirs at 700 °C and 50 °C. Determine its thermal efficiency, the net power output and the rate of heat rejection to the heat sink.

13.2 Calculate the required input power to a reversible refrigerator if it operates between thermal reservoirs at −15 °C and 35 °C and provides 12.6 kW of refrigeration. Determine also the rate of heat transfer to the heat sink.

13.3 An inventor has come up with a new concept: a device in which heat is transferred from the exhaust of a front-wheel-drive car and used to drive the rear wheels. The pressure of the exhaust stream entering the device is only slightly higher than the pressure of the environment. It is claimed that this device operates in a steady state accepting heat transfer from the stream of exhaust gas and rejecting heat to the environment. The inventor claims it can provide 9.7 kW of power when the exhaust flow rate is 0.048 kg/s, the exhaust temperature is 349 °C and ambient temperature is 17.6 °C. Assume that for the exhaust gases c_p = 1.005 kJ/kg K.

(a) Determine the maximum rate at which heat can be transferred from the exhaust to the steady state device.
(b) Determine the maximum possible thermal efficiency of any engine operating between a thermal reservoir at the stated exhaust temperature at entry to the device and a thermal reservoir at ambient temperature.
(c) Comment on the claims made by the inventor. Do they violate the first law of thermodynamics? Do they violate the second law of thermodynamics?

13.4 Show that if a system exists that violates the Clausius statement of the second law of thermodynamics, then the Kelvin–Planck statement is also violated.

13.5 Show that if the Kelvin–Planck statement of the second law is true, then the Clausius statement must also be true.

13.6 (a) Show how the Clausius inequality for a system that interacts with one thermal reservoir can yield the Kelvin–Planck statement of the second law.
(b) Show how the Clausius inequality for a system that interacts with two thermal reservoirs can yield the Clausius statement of the second law.
(c) Show how the Clausius inequality for a system that undergoes or contains only reversible processes, interacts with two thermal reservoirs, provides a net work output, has no other net effects on the surroundings and undergoes no net change can yield an expression for the thermal efficiency of this system in terms of the temperatures of the reservoirs.

CHAPTER 14

Entropy

A thermodynamic property called entropy is introduced and explained in this chapter. It is shown from the inequality of Clausius that there is a property whose change can be defined in terms of heat transfer and temperature for a reversible process between specified states. The property is given the name entropy. Where only entropy changes or differences are of interest, as in this book, the property *specific entropy* can be assigned an arbitrary value at some reference equilibrium state of a substance. Expressions that allow the entropy differences between various equilibrium states to be calculated are presented and explained. A short description of the physical significance of entropy is given to support its theoretical development. The temperature versus specific entropy equilibrium phase diagram for a pure substance is described. Entropy transfer and entropy transport are introduced. A general definition of work is given, which depends on the property entropy. The principle of increase of entropy is stated and proved from the inequality of Clausius. The relationships between the property entropy and the concept of an equilibrium state are discussed. The entropy balance equation for a system is described. A brief introduction is given to an area of thermodynamics known as exergy analysis.

Just as the first law of thermodynamics gives rise to the definition of a difference in the property internal energy, the second law gives rise to the definition of a difference in the property entropy. Both internal energy and entropy are extensive properties that cannot be measured directly. The first law provides the basis for the indirect measurement of internal energy changes or differences, while the second law and the first law provide the basis for the indirect measurement of entropy changes or differences. Because the mathematical form of the second law of thermodynamics involves an inequality for all cycles that are not reversible, it is found that the property entropy is not conserved: it can be created but not destroyed. These are complex issues and it is not easy to develop an intuitive grasp of the property entropy. For this reason it is particularly important to pay attention to the steps that lead to the definition of entropy differences between states. Each of these steps can readily be followed and none of them present major conceptual difficulties. As the property entropy is founded on the second law of thermodynamics, a clear understanding of Chapter 13 is advisable before proceeding further with this chapter.

14.1 The basis of entropy

14.1.1 The inequality of Clausius

The inequality of Clausius, which is an equality in the case of a reversible cycle, provides a convenient basis for establishing the existence of the property called entropy and for the definition of an entropy difference between states. The cyclic integral form for a reversible cycle, Equation (13.25), Chapter 13, will be used:

$$\oint \frac{\mathrm{d}Q}{T} = 0 \quad \text{for a reversible cycle.} \qquad (13.25, \text{Chapter } 13)$$

14.1.2 The existence of the property entropy

The existence of a property, which is called entropy, is explained with reference to Figure 14.1. Let A, B and C be reversible processes of a closed system.

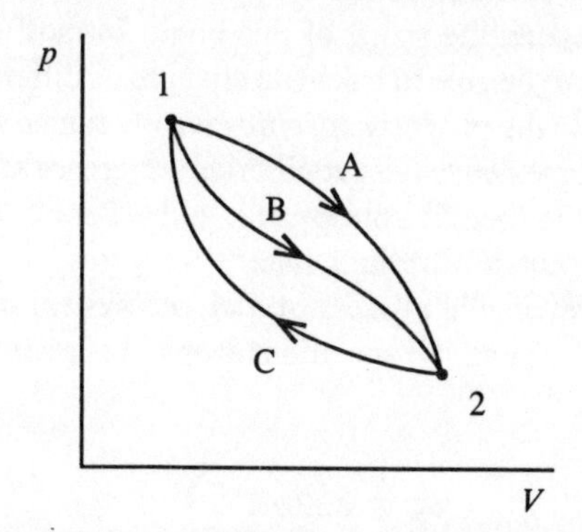

Figure 14.1 *Three reversible processes of a closed system between states 1 and 2.*

For a reversible cycle

$$\oint \frac{\mathrm{d}Q_{\mathrm{rev}}}{T} = 0.$$

For cycle AC

$$\underset{\mathrm{A}}{\int_1^2} \frac{\mathrm{d}Q}{T} + \underset{\mathrm{C}}{\int_2^1} \frac{\mathrm{d}Q}{T} = 0. \qquad (14.1)$$

For cycle BC

$$\underset{\mathrm{B}}{\int_1^2} \frac{\mathrm{d}Q}{T} + \underset{\mathrm{C}}{\int_2^1} \frac{\mathrm{d}Q}{T} = 0. \qquad (14.2)$$

Subtracting,

$$\int_{1 \atop A}^{2} \frac{dQ}{T} = \int_{1 \atop B}^{2} \frac{dQ}{T}. \tag{14.3}$$

Therefore, $\int_1^2 dQ_{rev}/T$ has the same value for all reversible processes from state 1 to state 2, irrespective of the process path. It therefore represents a difference in a point function of the independent properties that define the state (such as p and V) and is a property difference of the system between the two states. This property is given the name entropy.

14.2 The definition of entropy differences

The determination of absolute values of entropy depends on the *third law of thermodynamics*,[17] which is beyond the scope of this book. In most engineering applications it is only necessary to be able to calculate changes or differences in the property entropy. If this is the case the property *specific entropy* can be assigned an arbitrary value (usually zero) at some convenient equilibrium reference state of each substance involved. For instance, the specific entropy of liquid water can be assigned the value 0 J/kg K at the triple point equilibrium state.

For an infinitesimal reversible process of a closed system the entropy change is defined by Equation (14.4) and the specific entropy change is defined by Equation (14.5):

$$\delta S \triangleq \frac{\delta Q_{rev}}{T} \tag{14.4}$$

$$\delta s \triangleq \frac{\delta q_{rev}}{T}. \tag{14.5}$$

In order to find the difference in entropy between two states of a closed system, Equation (14.4) must be integrated for a reversible process between the states. It is unimportant which reversible process is chosen, as the value of the integral will be the same for all reversible processes that link the states (from the *equality* of Clausius). Therefore, for a reversible process between two states,

$$S_2 - S_1 \triangleq \int_1^2 \frac{dQ_{rev}}{T} \tag{14.6}$$

17. Roughly, this states that the absolute entropy of a pure substance can be assigned the value zero at absolute zero temperature.

Table 14.1

Property	*Symbol*	*Unit*
Entropy	S	J/K
Specific entropy	s	J/kg K

or, in intensive form,

$$s_2 - s_1 \triangleq \int_1^2 \frac{\mathrm{d}q_{\mathrm{rev}}}{T}. \tag{14.7}$$

Equations (14.6) and (14.7) allow the entropy and the specific entropy respectively at any state of a closed system to be determined once the entropy or specific entropy is known or has been assigned a reference value at some particular equilibrium state. The symbols and units for entropy and specific entropy are presented in Table 14.1.

The *entropy difference between two states* of a closed system is defined as the integral of the ratio of the incremental heat transfer to the absolute temperature for any reversible process between the states.

By considering a point system that passes through a steady flow open system, Equations (14.5) and (14.7) can be applied to a reversible steady flow process in which a substance passes from one equilibrium state to another with no change in composition. Equation (14.7) also allows the specific entropy of compounds and mixtures to be evaluated when the specific entropy values of the constituent substances are known or have been assigned values at specified equilibrium reference states – this is an important aspect of chemical thermodynamics, but is beyond the scope of this book.

14.3 Entropy changes for various processes

14.3.1 Reversible isothermal process

$$\Delta T_{1\to2} = 0.$$

The entropy change is given by

$$\Delta S_{1\to2} = \int_1^2 \frac{\mathrm{d}Q_{\mathrm{rev}}}{T} = \frac{1}{T}\int_1^2 \mathrm{d}Q_{\mathrm{rev}} = \frac{Q_{1\to2}}{T}. \tag{14.8}$$

Also

$$\Delta s_{1\to2} = \frac{q_{1\to2}}{T}. \tag{14.9}$$

Flow processes where a phase change occurs at constant pressure in boilers, evaporators and condensers are examples of practical isothermal processes.

14.3.2 Reversible adiabatic process

For any part of the process

$$\delta S = \frac{\delta Q_{rev}}{T},$$

but

$$\delta Q_{rev} = 0.$$

Therefore

$$\Delta S_{1\to 2} = \Sigma \delta S = 0. \tag{14.10}$$

Also

$$\Delta s_{1\to 2} = 0. \tag{14.11}$$

An *isentropic process* is any process during which the entropy remains constant. From Equation (14.4), all reversible adiabatic processes are isentropic and all reversible isentropic processes are adiabatic.

An ideal reversible adiabatic compressor and an ideal reversible adiabatic nozzle are two examples of devices in which a reversible adiabatic process would occur. Real processes are never perfectly reversible or perfectly adiabatic.

14.3.3 Reversible constant pressure process (non-flow, ideal gas)

$$\Delta p_{1\to 2} = 0.$$

The non-flow energy equation is written as

$$Q + W = \Delta U.$$

Hence

$$Q = \Delta U - W = \Delta U + p\Delta V.$$

But, as p is constant,

$$p\Delta V = \Delta(pV)$$

and so

$$Q = \Delta U + \Delta(pV) = \Delta(U + pV) = \Delta H.$$

For an infinitesimal change

$$\mathrm{d}Q = \mathrm{d}H = mc_p\,\mathrm{d}T.$$

$$\Delta S_{1\to2} = \int_1^2 \frac{\mathrm{d}Q_{\text{rev}}}{T} = \int_1^2 \frac{mc_p\,\mathrm{d}T}{T} = m\int_1^2 \frac{c_p\,\mathrm{d}T}{T}. \tag{14.12}$$

If c_p is constant

$$\Delta S_{1\to2} = mc_p \int_1^2 \frac{\mathrm{d}T}{T} = mc_p \ln\left(\frac{T_2}{T_1}\right). \tag{14.13}$$

Equation (14.13) can be used, for instance, to calculate the change in entropy between two states of a closed system containing an ideal gas, where the states have the same pressure but different temperatures.

14.3.4 Reversible constant volume process (non-flow, ideal gas)

$$\Delta V_{1\to2} = 0.$$

The non-flow energy equation is written as

$$Q + W = \Delta U.$$

But there is no work as the volume is constant. Therefore

$$Q = \Delta U.$$

For an infinitesimal change

$$\mathrm{d}Q = \mathrm{d}U = mc_v\,\mathrm{d}T.$$

$$\Delta S_{1\to2} = \int_1^2 \frac{\mathrm{d}Q_{\text{rev}}}{T} = \int_1^2 \frac{mc_v\,\mathrm{d}T}{T} = m\int_1^2 \frac{c_v\,\mathrm{d}T}{T}. \tag{14.14}$$

If c_v is constant

$$\Delta S_{1\to2} = mc_v \int_1^2 \frac{\mathrm{d}T}{T} = mc_v \ln\left(\frac{T_2}{T_1}\right). \tag{14.15}$$

EXAMPLE 14.1

Calculate the change in entropy of a closed system containing 3.29 litres of argon gas in equilibrium at 20 °C and 0.11 MPa when it undergoes a process at the end of which it is again in equilibrium, but at a temperature of 124 °C and a volume of 2.92 litres. For argon, R = 0.2081 kJ/kg K, c_p = 0.5203 kJ/kg K, c_v = 0.3122 kJ/kg K and γ = 1.667.

SOLUTION

As entropy is a property, the entropy change for a process between states is independent of the process path. To evaluate the change it is convenient to consider a process path made up of two reversible processes where a property is held constant for each of these. Suppose the system undergoes first a reversible isothermal process and then a reversible constant volume process, as shown in Figure 14.2. Let A refer to the state after the isothermal process.

$$Q_{1\to A} = -W_{1\to A} = -p_1 V_1 \ln\frac{p_A}{p_1} = -p_1 V_1 \ln\frac{V_1}{V_A}$$

$$= -0.11 \times 10^6\ [\mathrm{N\ m^{-2}}]\ 3.29 \times 10^{-3}\ [\mathrm{m^3}] \ln\left(\frac{3.29\ [\mathrm{L}]}{2.92\ [\mathrm{L}]}\right)$$

$$= -43.18\ \mathrm{J}$$

$$\Delta S_{1\to A} = \frac{Q_{1\to A}}{T_1} = \frac{-43.18\ [\mathrm{J}]}{20\ [°\mathrm{C}] + 273.15\ [\mathrm{K}]} = -0.1473\ \mathrm{J/K}$$

$$m = \frac{p_1 V_1}{R T_1} = \frac{0.11 \times 10^6\ [\mathrm{N\ m^{-2}}]\ 3.29 \times 10^{-3}\ [\mathrm{m^3}]}{208.1\ [\mathrm{J\ kg^{-1}\ K^{-1}}]\ 293.15\ [\mathrm{K}]} = 5.932 \times 10^{-3}\ \mathrm{kg}$$

$$\Delta S_{A\to 2} = m c_v \ln\left(\frac{T_2}{T_A}\right)$$

$$= 5.932 \times 10^{-3}\ [\mathrm{kg}]\ 312.2\ [\mathrm{J\ kg^{-1}\ K^{-1}}] \ln\left(\frac{124\ [°\mathrm{C}] + 273.15\ [\mathrm{K}]}{20\ [°\mathrm{C}] + 273.15\ [\mathrm{K}]}\right)$$

$$= 0.5623\ \mathrm{J/K}$$

$$\Delta S_{1\to 2} = \Delta S_{1\to A} + \Delta S_{A\to 2}$$

$$= (-0.1473 + 0.5623)\ \mathrm{J/K}$$

Answer

$$= 0.415\ \mathrm{J/K}.$$

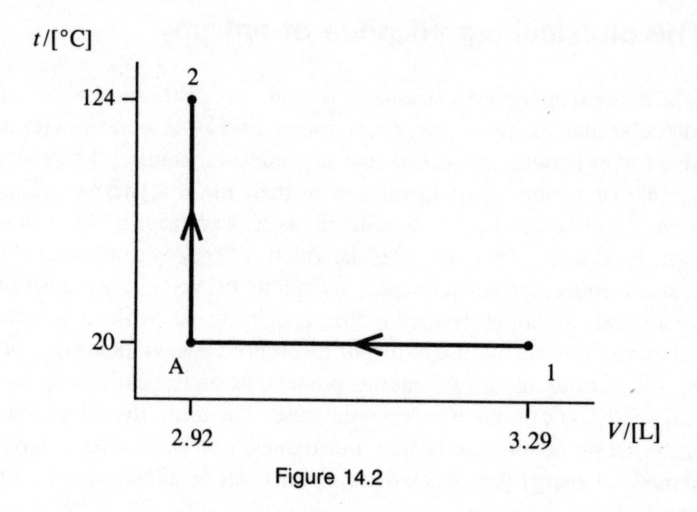

Figure 14.2

By following the type of procedure used in Example 14.1, the entropy difference between any two equilibrium states can be determined if sufficient information is available to describe the states. From the state proposition only two independent thermodynamic properties are required to specify the equilibrium thermodynamic state of a compressible homogeneous substance. Therefore, for such a substance, the initial and final states can be specified by giving the initial and final values of two properties. The states can always be linked by a reversible process where one of the properties is held constant, followed by a reversible process where the other property is held constant.

14.4 Heat transfer as a path function for a reversible process

For an infinitesimal reversible process of a closed system, heat transfer can be expressed as a function of the entropy change and the absolute temperature by rearranging Equation (14.4):

$$\delta Q_{\text{rev}} = T\delta S. \tag{14.16}$$

For a finite reversible process the heat transfer is given by

$$Q_{\text{rev}} = \int_1^2 T\,\mathrm{d}S \tag{14.17}$$

or, per unit mass,

$$q_{\text{rev}} = \int_1^2 T\,\mathrm{d}s. \tag{14.18}$$

Equations (14.17) and (14.18) describe heat transfer as a path function of equilibrium properties for a reversible process.

14.5 The physical significance of entropy

Much of the internal energy of a substance is randomly distributed as kinetic energy at the molecular and submolecular levels and as energy associated with attractive or repulsive forces between molecular and submolecular entities, which are moving closer together or further apart in relation to their mean separation. This energy is sometimes described as being 'disordered' as it is not accessible as work at the macroscopic level in the same way as is the kinetic energy or gravitational potential energy that an entire system possesses owing to its velocity or position in the gravitational field. Although energy is the capacity to do work, it is not possible directly to access the minute quantities of disordered energy possessed at a given instant by the various modes of energy possession of the entities so as to yield mechanical shaft work on the macroscopic scale. The term 'disorder' refers to the lack of information about exactly how much energy is associated at any moment with each mode of energy possession of each molecular or submolecular entity within the system.

At the molecular and submolecular level there is also 'ordered energy' associated with attractive or repulsive forces between entities that have fixed mean relative positions. Part of this energy is, in principle, accessible as work at the macroscopic level under very special conditions, which are beyond the scope of this book.

Temperature is the property that determines whether a system that is in equilibrium will experience any decrease or increase in its disordered energy if it is brought into contact with another system that is in equilibrium. If the systems do not have the same temperature, disordered energy will be redistributed from the system at the higher temperature to the one at the lower temperature. There is then less information about precisely where that energy resides, as it is now dispersed over the two systems.

Heat transfer to a system increases the disordered energy of the system. Heat transfer from a system reduces the disordered energy. Reversible heat transfer is characterized by both the amount of energy transferred to or from the system and the temperature level at which this occurs. The property entropy, whose change between states is defined as the integral of the ratio of the reversible heat transfer to the absolute temperature, is a measure of the state of disorder of the system. This 'state of disorder' is characterized by the amount of disordered energy and its temperature level. When reversible heat transfer occurs from one system to another, both systems have the same temperature and the increase in the disorder of one is exactly matched by the decrease in disorder of the other.

When reversible adiabatic work is done on or by a system its ordered energy increases or decreases by exactly the amount of the work and the temperature level changes in a way that depends on the substances involved. Reversible work is characterized by the amount of energy transferred to or from the system, irrespective of the temperature of the system.

Irreversible work, such as stirring work or friction work between subsystems, involves a change in the disorder of the system and, like heat transfer to a system, has the effect of increasing the entropy.

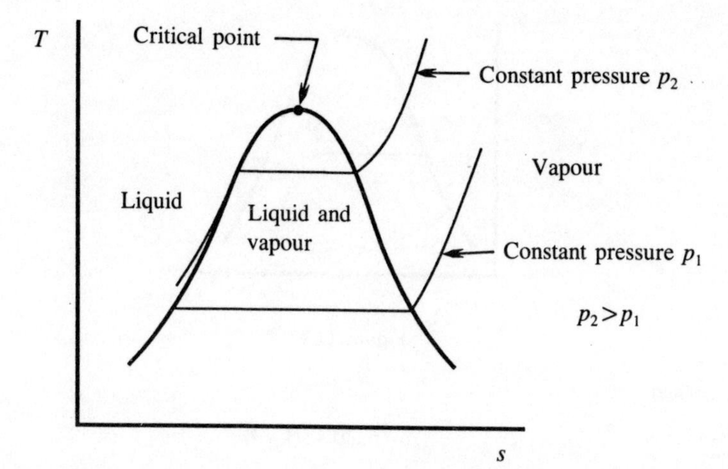

Figure 14.3 *Temperature versus specific entropy equilibrium diagram for a pure substance.*

14.6 The temperature versus specific entropy diagram

The specific entropy at all equilibrium states can be calculated, given the entropy at some reference state, using equations such as those that have been described for reversible processes between equilibrium states. For the liquid and vapour states of a pure substance, a temperature versus specific entropy equilibrium diagram as shown in Figure 14.3 can be drawn.

EXAMPLE 14.2
Determine the increase in specific entropy of saturated liquid water at a temperature of 100 °C when it is evaporated at constant pressure to wet vapour with a dryness fraction of 0.86.

(1) Use the specific entropy values directly. (2) Use the relationship

$$\Delta s_{1\to 2} = \int_1^2 \frac{\mathrm{d}q_{\mathrm{rev}}}{T}.$$

SOLUTION (see Figure 14.4)
Method 1: From the tables

$$s_1 = s_{\mathrm{f}} = 1.307 \text{ kJ/kg K}$$

$$s_2 = s_{\mathrm{f}} + x_2 s_{\mathrm{fg}} = 1.307 + 0.86(6.049) \text{ kJ/kg K}$$

$$= 6.509 \text{ kJ/kg K}$$

$$\Delta s_{1\to 2} = s_2 - s_1 = 6.509 - 1.307 \text{ kJ/kg K}$$

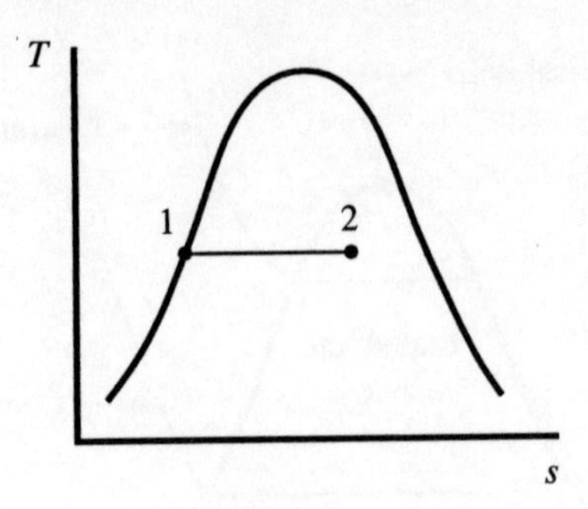

Figure 14.4

Answer

$$= 5.20 \text{ kJ/kg K}.$$

Method 2: Assume a reversible constant pressure process 1→2

$$q = \Delta u - w$$

$$q = (u_2 - u_1) + p(v_2 - v_1)$$

$$= h_2 - h_1.$$

From the tables

$$h_f = 419.1 \text{ kJ/kg}$$

$$h_{fg} = 2256.9 \text{ kJ/kg}$$

$$h_1 = h_f = 419.1 \text{ kJ/kg}$$

$$h_2 = h_f + x_2 h_{fg} = 419.1 + 0.86(2256.9) \text{ kJ/kg}$$

$$= 419.1 + 1940.9 \text{ kJ/kg}$$

$$= 2360.0 \text{ kJ/kg}.$$

For the constant pressure process of a two-phase pure substance, the temperature is constant:

$$T = 373.15 \text{ K}$$

$$\Delta s_{1\rightarrow 2} = \frac{q}{T} = \frac{h_2 - h_1}{T} = \frac{2360.0 - 419.1}{373.15} \text{ kJ/kg K}$$

Answer

$$= 5.20 \text{ kJ/kg K}.$$

It is seen that both methods give the same result.

14.7 Entropy transfer and transport

Like the property internal energy, entropy can be transported or transferred across the boundary of a system. Entropy transfer occurs only in association with heat transfer. Entropy transport occurs in association with mass crossing a boundary.

Figure 14.5 shows a patch, perhaps of nearly infinitesimal area, on the impermeable boundary of a system. Energy crosses the patch as heat transfer Q_{in} and non-heat energy transfer W_{in}. The entropy transfer across the boundary of a system in association with a minute quantity of energy transfer consisting of heat transfer δQ_{in} and non-heat energy transfer δW_{in} over a minute area of a system boundary, which is at a temperature T_{bdry}, is defined by Equation (14.19).

$$\delta S_{\text{in},Q+W} = \delta S_{\text{in},Q} \stackrel{\Delta}{=} \frac{\delta Q_{\text{in}}}{T_{\text{bdry}}}. \tag{14.19}$$

In general, the amount of heat transfer per unit area and the temperature may both vary over the boundary surface and with time. Therefore, in order to evaluate the entropy transfer into a system, it is necessary to sum up the amounts of entropy transfer for a very large number of minute areas and for a very large number of minute time intervals. For the relatively simple case of heat transfer at a boundary

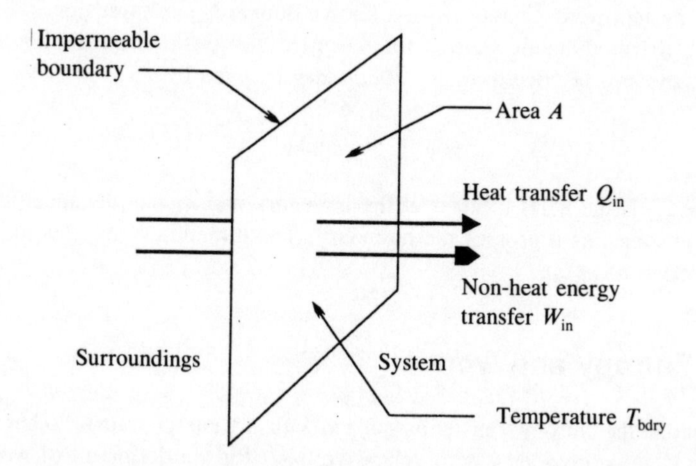

Figure 14.5 *A patch on an impermeable boundary of a system where energy crosses as heat transfer and as non-heat energy transfer.*

that is at a constant temperature, the entropy transfer into the system is given by Equation (14.20):

$$S_{\text{in},Q} = \frac{Q_{\text{in}}}{T_{\text{bdry}}}. \tag{14.20}$$

Similar expressions to Equations (14.19) and (14.20) would apply for the entropy transfer out of a system. In general, the *entropy transfer at a boundary* is given by

$$\delta S_{\text{trnsf}} = \frac{\delta Q}{T_{\text{bdry}}} \tag{14.21}$$

where δS_{trnsf} has the same conventional positive direction as δQ: the inwards direction. The total entropy transfer across a boundary is given by

$$S_{\text{trnsf}} = \int_{\text{bdry}} \frac{\mathrm{d}Q}{T_{\text{bdry}}} \tag{14.22}$$

where the integration is performed over the surface area of the boundary.

The entropy transport across the boundary of an open system in association with a known inwards mass transfer and a known intensive thermodynamic state at the specified inlet position on the boundary is given by Equation (14.23):

$$S_{\text{in},m} = m_{\text{in}} s_{\text{in}}. \tag{14.23}$$

A similar expression would apply for the entropy transport across the boundary of an open system in association with a known outwards mass transfer and a known intensive thermodynamic state at the specified exit position on the boundary. In general, the *entropy transport at a boundary* is given by

$$S_{\text{trnsp}} = m_{\text{trnsf}} s_{\text{bdry}} \tag{14.24}$$

where m_{trnsf} is the mass transfer at the boundary and s_{bdry} is the specific entropy of the substance as it crosses the boundary. The direction of S_{trnsp} is the same as the direction of m_{trnsf}.

14.8 Entropy and work

The fact that the entropy transfer associated with any energy transfer other than heat transfer is zero provides a comprehensive basis for the definition of work. Until now only mechanical work has been specifically defined. *Work* is energy transfer across a boundary that has no associated entropy transfer. All forms of energy transfer across a boundary that do not have associated entropy transfer or substance transfer are equivalent to mechanical shear work in terms of the first and second laws of thermodynamics. Power transmitted in an electric cable is an example of a rate of non-mechanical work. Work can also occur owing to a changing magnetic field. The power transmission from the primary to the secondary winding of a transformer or from the stator of an induction motor to the rotor involves this type of work.

14.9 The principle of increase of entropy

The entropy of an isolated system can only increase and cannot decrease. This can be shown from the inequality of Clausius, as follows. Suppose an isolated system at state 1 undergoes changes, described as process A, after which it is at state 2. Now consider a reversible process, B, in which the system remains closed but no longer isolated, that would bring the system back to its original state. The original process A from state 1 to state 2 and the reversible process B from state 2 to state 1 constitute a cycle of a closed system and the Clausius inequality applies:

$$\int_1^2 \left(\frac{\mathrm{d}Q}{T}\right)_{\mathrm{A}} + \int_2^1 \left(\frac{\mathrm{d}Q}{T}\right)_{\mathrm{B,rev}} \le 0. \tag{14.25}$$

But as the system is isolated for process A, there is no heat transfer across the boundary and so

$$\int_1^2 \left(\frac{\mathrm{d}Q}{T}\right)_{\mathrm{A}} = 0. \tag{14.26}$$

Also, from the definition of an entropy change,

$$\int_2^1 \left(\frac{\mathrm{d}Q}{T}\right)_{\mathrm{B,rev}} = S_1 - S_2. \tag{14.27}$$

Combining Equations (14.26) and (14.27) with the inequality, Equation (14.25),

$$S_1 - S_2 \le 0. \tag{14.28}$$

Therefore

$$S_2 \ge S_1. \tag{14.29}$$

Thus, when the state of an isolated system changes, the entropy after the change is greater than or equal to the entropy before the change.

14.9.1 Entropy and stable equilibrium

A system is said to be in a *stable equilibrium state* if it can undergo no change without there being a net effect on its surroundings. The existence of a unique state of stable equilibrium for any system, defined in terms of thermodynamic properties, is a consequence of the second law of thermodynamics. In fact, the statement that *there exists a unique stable equilibrium state for any system at any moment in time* is itself a statement of the second law of thermodynamics.[18] Further explanation of this particular statement of the second law is beyond the scope of this textbook.

In certain cases a system may be in an equilibrium state, but a relatively minor disturbance could cause a very significant change in its state. A system is said to be in an *unstable equilibrium state* if no change would occur if it were isolated, but where a non-permanent change in the surroundings (such as a temporary transfer of energy from the surroundings as heat or work) could cause a significant change in the state of the system while it was otherwise isolated.

Equation (14.29), which states that the entropy increases or remains unchanged in any process of an isolated system, applies as an equality only if the process is reversible. Therefore, if there is no change in the entropy of an isolated system, all the processes that occur within it must be reversible. It is known from everyday experience that an isolated system does not continue to change indefinitely, but comes to a state of no change, which is an equilibrium state. The system does not continue to undergo reversible changes because all real processes involve mechanisms such as friction, or fluid friction, or heat transfer with finite temperature differences, which are irreversible. Therefore, in practice the entropy of an isolated system reaches a maximum at equilibrium, when all changes cease.

The reversible work and the reversible heat transfer for any part of an equilibrium process can be uniquely expressed as a function of the thermodynamic properties along the process path (for instance, Equation (3.16) of Chapter 3, $W = -\int p \, dV$, and Equation (14.17), $Q_{rev} = \int_1^2 T \, dS$). If the process were reversed, the values of the heat transfer and the work for the reverse process would be opposite in sign to those for the forward process. The net heat transfer and the net work would thus be zero when the process had been reversed along the original path. Therefore, *an equilibrium process is a special case of a reversible process.*

14.9.2 Increasing total entropy

The surroundings of any system are infinite as they include all systems and all space that are not part of the system. The following corollary to the principle of increase of entropy is valid if the composite of a system and its surroundings can be considered to be enclosed by an isolating boundary. *Whenever an irreversible process occurs within a system, there is an increase in the total entropy of the system and its surroundings.* This follows from the principle of increase of entropy as any irrevers-

18. A statement of this type has been put forward by Gyftopoulos and Beretta (1991). See Appendix D.

ibility within the composite of the system and its surroundings is sufficient to cause an increase in the entropy of the composite system:

$$\Delta S_{sys} + \Delta S_{surr} \geq 0. \qquad (14.30)$$

The entropy changes ΔS_{sys} and ΔS_{surr} in Equation (14.30) are then taken to be those associated with a particular process of the system. For practical purposes, the surroundings can be considered to be enclosed by a boundary at such a distance that there would be no effect there of changes undergone by the system.

The total entropy of a system and its surroundings can only remain unchanged if the system and its surroundings undergo reversible processes. Entropy is always created whenever an irreversible process occurs. There is no known mechanism by which entropy can be destroyed.

14.10 The entropy balance equation

An entropy balance equation, which is somewhat analogous to a mass or energy balance equation, can be written for any system. However, in the case of the property entropy, an extra term is needed to take account of the fact that entropy is not conserved. This is the entropy creation term S_{cr}. Equation (14.31) is the entropy balance equation. An entropy balance can also be written as a time rate equation as in Equation (14.32):

$$S_{in} + S_{cr} = S_{out} + \Delta S \qquad (14.31)$$

$$\dot{S}_{in} + \dot{S}_{cf} = \dot{S}_{out} + \dot{S}. \qquad (14.32)$$

EXAMPLE 14.3

The mean temperature of the external surface of a heating panel is 68.9 °C. Heat transfer occurs from the surface at the rate of 1.5 kW to the air in a room, which has a mean temperature of 21.7 °C (see Figure 14.6). The temperature drop between the surface of the heating panel and the main body of air occurs in a thin boundary layer of air at the surface of the panel. By using the two mean temperatures mentioned, calculate the rate of entropy creation in the boundary layer due to the irreversible heat transfer. As an approximation, it can be assumed that the boundary layer serves as a heat conductor and is in a steady state with no flow.

SOLUTION

As the boundary layer is in a steady state, the rate of change of its entropy is zero.

The rate of entropy transfer into the boundary layer is given by

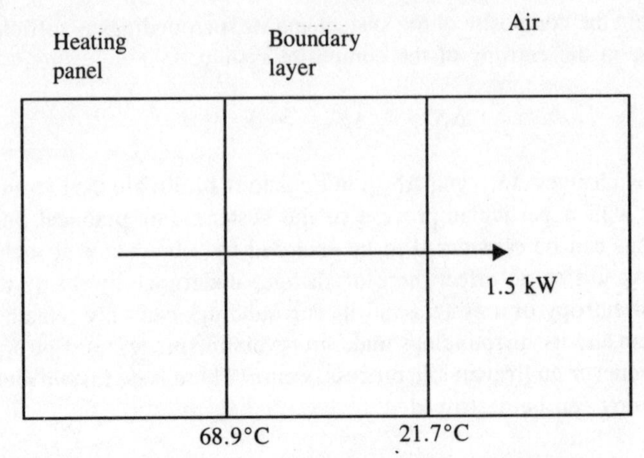

Figure 14.6 *Schematic representation of the heating panel, the heat transfer boundary layer and the air.*

$$\dot{S}_{in} = \frac{\dot{Q}_{in}}{T_{bdry}}$$

$$\dot{S}_{in} = \frac{1.5\ [\text{kW}]}{(68.9 + 273.15)\ [\text{K}]} = 4.39 \times 10^{-3}\ \text{kW/K}$$

$$\dot{S}_{out} = \frac{1.5\ [\text{kW}]}{(21.7 + 273.15)\ [\text{K}]} = 5.09 \times 10^{-3}\ \text{kW/K}$$

$$\dot{S}_{cr} = \dot{S} + \dot{S}_{out} - \dot{S}_{in} = (0 + 5.09 - 4.39)\ \text{W/K}$$

Answer

$$= 0.70\ \text{W/K}.$$

14.11 Exergy analysis

A very important class of problems in engineering thermodynamics concerns systems or substances that can be modelled as being in equilibrium or stable equilibrium, but that are not in mutual stable equilibrium with the surroundings. For instance, within the Earth there are reserves of fuels that are not in mutual stable equilibrium with the atmosphere and the sea. The requirements of mutual chemical equilibrium are not met. Any system at a temperature above or below that of the environment is not in mutual stable equilibrium with the environment. In this case the requirements of mutual thermal equilibrium are not met. It is found that any lack of mutual stable equilibrium between a system and the environment can be used to produce shaft work. The second law of thermodynamics allows the maximum work that could be produced to be determined.

The *exergy of a system* is defined as the maximum shaft work that could be done by the composite of the system and a specified reference environment that is assumed to be infinite, in equilibrium, and ultimately to enclose all other systems.

Typically, the environment is specified by stating its temperature, pressure and chemical composition. Exergy is not simply a thermodynamic property, but rather is a co-property of a system and the reference environment. The term *exergy* comes from the Greek words *ex* and *ergon*, meaning *from* and *work*: the exergy of a system can be increased if work is done on it. The following are some terms found in the literature that are equivalent to or nearly equivalent to *exergy*: available energy, essergy, utilizable energy, available work, availability.

Exergy has the characteristic that it is conserved only when all processes of the system and the environment are reversible. Exergy is destroyed whenever an irreversible process occurs. When an exergy analysis is performed on a plant such as an entire power station, a chemical processing plant or a refrigeration plant, the thermodynamic imperfections can be quantified as exergy destruction, which is wasted shaft work or wasted potential for the production of shaft work.

Like energy, exergy can be transferred or transported across the boundary of a system. For each type of energy transfer or transport there is a corresponding exergy transfer or transport. In particular, exergy analysis takes into account the different thermodynamic values of work and heat. The exergy transfer associated with shaft work is equal to the shaft work. The exergy transfer associated with heat transfer, however, depends on the temperature level at which it occurs in relation to the temperature of the environment, T_0. It is given by Equation (14.33):

$$\Xi_Q \triangleq \frac{T - T_0}{T} Q \tag{14.33}$$

where Ξ_Q (Ξ is pronounced 'Xi', as in xylophone) is the exergy transfer corresponding to the heat transfer. It can be noted that if the heat transfer occurs at a temperature below the temperature of the environment, the direction of the exergy transfer is opposite to that of the energy transfer. For example, a refrigerator that takes energy from a low-temperature system provides an exergy output to that system.

EXAMPLE 14.4

If a heat engine receives 100 kW of heat transfer from a thermal reservoir at 270 °C and rejects heat to the environment at 20 °C while producing shaft work at a rate of 15 kW, what is the rate of exergy input? What would be the rate of heat input and exergy input of a reversible heat engine that operated between the same thermal reservoirs and provided the same rate of work output?

SOLUTION

From Equation (14.33), there is no exergy transfer associated with heat rejection to the environment. The rate of exergy input corresponding to the heat input is given by

Answer

$$\dot{\Xi}_{Q,\text{in}} = \frac{T - T_0}{T}\dot{Q}_{\text{in}} = \frac{(270 - 20)\ [\text{K}]}{(270 + 273.15)\ [\text{K}]} 100\ [\text{kW}] = 46.0\ \text{kW}.$$

For a reversible heat engine the heat input would be given by

Answer

$$\dot{Q}_{\text{in,rev}} = \frac{\dot{W}_{\text{net,out}}}{\eta_{\text{th,rev}}} = 15\ [\text{kW}]\ \frac{T}{T - T_0} = \frac{(270 + 273.15)\ [\text{K}]}{(270 - 20)\ [\text{K}]} 15\ [\text{kW}]$$
$$= 32.59\ \text{kW}.$$

The exergy input of the reversible engine would be

Answer

$$\dot{\Xi}_{Q,\text{in,rev}} = \frac{T - T_0}{T}\dot{Q}_{\text{in,rev}} = \frac{(270 - 20)\ [\text{K}]}{(270 + 273.15)\ [\text{K}]} 32.59\ [\text{kW}] = 15.0\ \text{kW}.$$

14.12 Practical tips

- The temperature in the defining expression for an entropy change and in the expression for the entropy transfer to or from a system is an absolute temperature. Make sure that numerical values are in kelvin units.
- Entropy is an extensive thermodynamic property. Like the extensive properties mass and internal energy, it has a value at every state, whether or not the system is in equilibrium.
- For ideal gases, entropy changes or differences between different equilibrium states can be calculated readily using the types of expression that have been presented in this chapter. Also, for ideal gases, reversible isentropic processes arc described by the equation pV^γ = const or pv^γ = const.
- For substances that cannot be regarded as ideal gases, such as water substance in the vapour phase, thermodynamic tables are used to look up specific entropy values.

14.13 Summary

From the inequality of Clausius, which is an equality for a reversible cycle, it has been shown that $\int_1^2 \mathrm{d}Q_{rev}/T$ represents a property difference between states 1 and 2. This property is given the name entropy. A definition in words for the entropy difference between two states has also been given. It has been pointed out that the property specific entropy can be assigned an arbitrary value at some convenient equilibrium state of a particular substance. Explanations have been given of how the entropy change between equilibrium states can be evaluated for a reversible isothermal process, a reversible adiabatic process, a reversible constant pressure process of a closed system that contains an ideal gas and a reversible constant volume process of a closed system that contains an ideal gas. It has also been indicated how entropy differences between arbitrary states can be determined. The physical significance of entropy has been discussed. The temperature versus specific entropy equilibrium phase diagram for a pure substance has been described. Entropy transfer and transport have been explained and expressions given for these quantities. A general definition of work has been given, based on the concept of entropy. The principle of increase of entropy has been stated and proved. Stable equilibrium has been defined. The statement of increasing total entropy has been given as a corollary to the principle of increase of entropy. The entropy balance equation has been described. A very brief introduction to exergy analysis has been provided.

The student should be able to

- *define*
  — the change in entropy for an infinitesimal reversible process, the change in entropy for a reversible process between two equilibrium states, the difference in entropy between two states of a closed system, a reversible process, entropy transfer, entropy transport, work, a stable equilibrium state, an unstable equilibrium state, the exergy of a system
- *sketch*
  — the temperature versus specific entropy equilibrium phase diagram for a pure substance
- *show*
  — that $\int_1^2 \mathrm{d}Q_{rev}/T$ represents a property difference between states 1 and 2
- *explain*
  — the sequence of steps that leads from the inequality of Clausius to the definition of a difference in the property entropy, the physical significance of entropy, the statement of increasing total entropy, exergy analysis

- *derive*
  — expressions for the entropy change for: a reversible isothermal process, a reversible adiabatic process, a reversible constant pressure process of a closed system that contains an ideal gas, a reversible constant volume process of a closed system that contains an ideal gas
- *prove*
  — the principle of increase of entropy
- *state*
  — the units of entropy, the units of specific entropy, the principle of increase of entropy
- *write*
  — the entropy balance equation for a system
- *use*
  — the steam tables to look up values of the specific entropy
- *calculate*
  — entropy changes between equilibrium states by using fundamental relationships or entropy values from thermodynamic tables, quantities or rates of entropy transfer or transport.

14.14 Self-assessment questions

14.1 By considering a reversible isentropic process followed by a reversible constant pressure process, evaluate the overall entropy change for a closed system that initially contains 3.29 litres of argon gas at 20 °C and 0.11 MPa and finally reaches an equilibrium state at which its temperature is 124 °C and its volume is 2.92 litres. For argon, R = 0.2081 kJ/kg K, c_p = 0.5203 kJ/kg K, c_v = 0.3122 kJ/kg K and γ = 1.667. The end states are the same as for Example 14.1 in this chapter.

14.2 Derive an expression for the specific entropy change in terms of end-state pressures and temperatures, the specific heat at constant pressure and the specific gas constant for any process of an ideal gas between equilibrium states. Assume the specific heat capacities are constant.

14.3 Use the steam tables to find:
(a) The specific entropy of liquid water at 7 °C.
(b) The specific entropy of wet steam with a dryness fraction of 0.9 and a pressure of 1 MPa.
(c) The specific entropy of steam that is at a temperature of 400 °C and a pressure of 10 MPa.

CHAPTER 15

The Carnot cycle

The Carnot cycle is introduced in this chapter. This is a reversible cycle that was proposed by Sadi Carnot in 1824 for an ideal heat engine that would operate between two thermal reservoirs and a mechanical reservoir. It is described for an ideal gas working fluid and also for a two-phase, liquid and vapour, working fluid.

The Carnot cycle is not used directly as the basis of real engines, owing to practical difficulties in implementing it. However, some actual engine cycles are related to the Carnot cycle and share some of its features.

15.1 Description of the Carnot cycle

Carnot proposed a reversible cycle for an ideal heat engine that would operate between two thermal reservoirs and a mechanical reservoir. It consists of four equilibrium processes:

$1 \rightarrow 2$ isothermal heat rejection
$2 \rightarrow 3$ adiabatic temperature increase
$3 \rightarrow 4$ isothermal heat acceptance
$4 \rightarrow 1$ adiabatic temperature decrease.

Various working fluids could be used; for example, water and steam (two phase) or a gas such as air. A reversible Carnot cycle engine has the same thermal efficiency as all reversible engines that operate between thermal reservoirs at T_H and T_L (Equation (13.9), Chapter 13). This thermal efficiency is known as the *Carnot efficiency*:

$$\eta_{\text{Carnot}} = \eta_{\text{th,rev}} = \frac{T_H - T_L}{T_H}. \quad (15.1)$$

In Figure 15.1, the Carnot cycle is shown on a T–s diagram. The cycle has a rectangular shape on this diagram no matter what working fluid is used. The reversible adiabatic temperature-increase process $2 \rightarrow 3$ and the reversible adiabatic temperature-decrease process $4 \rightarrow 1$ are isentropic processes and so are vertical lines on the T–s diagram. The isothermal processes are horizontal lines.

The thermal efficiency of the Carnot cycle can be derived very quickly from the

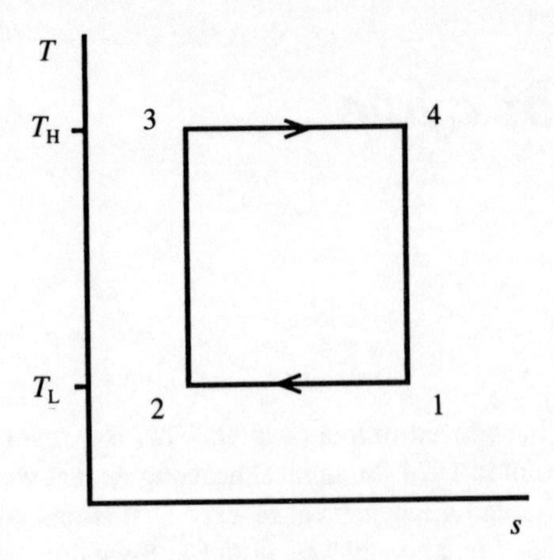

Figure 15.1 *The Carnot cycle.*

shape of the cycle on the T–s diagram, as follows. Equation (14.9), Chapter 14, relates the entropy change to the heat transfer and the temperature for a reversible isothermal process. This can be applied to process 1→2 of the Carnot cycle:

$$\Delta s_{1\to2} = s_2 - s_1 = \frac{q_{1\to2}}{T_L}. \tag{15.2}$$

Rearranging,

$$q_{1\to2} = T_L(s_2 - s_1). \tag{15.3}$$

As $s_2 < s_1$, $q_{1\to2}$ is negative and represents heat rejection from the system and so

$$q_{L,out} = -q_{1\to2} = T_L(s_1 - s_2). \tag{15.4}$$

By similar reasoning

$$q_{H,in} = q_{3\to4} = T_H(s_4 - s_3). \tag{15.5}$$

The thermal efficiency is given by

$$\eta_{Carnot} = \frac{q_{H,in} - q_{L,out}}{q_{H,in}} = \frac{T_H(s_4 - s_3) - T_L(s_1 - s_2)}{T_H(s_4 - s_3)}. \tag{15.6}$$

As $s_4 = s_1$ and $s_3 = s_2$,

$$\eta_{Carnot} = \frac{T_H - T_L}{T_H}. \qquad \text{(Equation 15.1)}$$

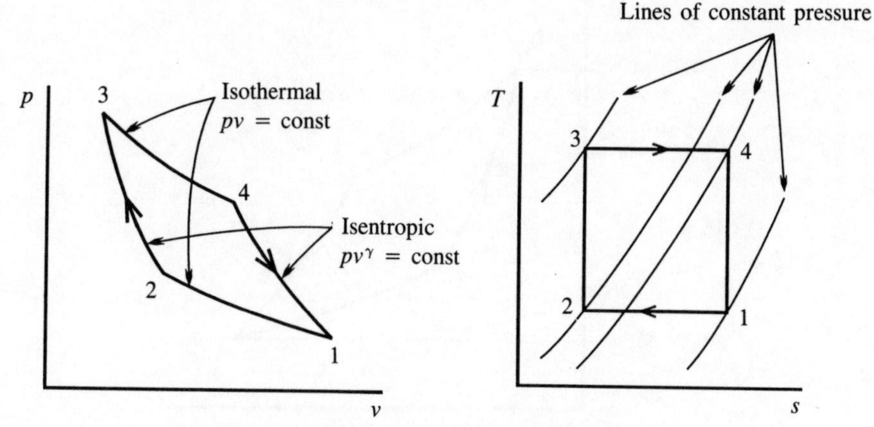

Figure 15.2 *The Carnot cycle for an ideal gas.*

15.2 The Carnot cycle for an ideal gas

Figure 15.2 illustrates the Carnot cycle for an ideal gas. This cycle could be undergone by a non-flow system or could be contained within a steady flow system. The cycle will be described here for a non-flow system.

In the first process the temperature remains constant while there is heat rejection from the system. This can only occur if the volume of the gas reduces. The process can therefore be described as isothermal compression. In the second process there is no heat transfer while the temperature rises. In this case too, the gas volume reduces. The process can be described as adiabatic compression. The work is positive for the first two processes.

In the third process there is heat transfer to the system while the temperature remains constant. This can only happen if the volume of the gas increases. The process can be described as isothermal expansion. In the last process there is no heat transfer while the temperature of the gas reduces to the original value at the start of the cycle and the gas volume increases to its original value. This process can be described as adiabatic expansion. The work is negative for the last two processes.

EXAMPLE 15.1

Case study: specification of an ideal non-flow Carnot gas cycle

Specify and analyze a Carnot cycle to operate with a heat source at 350 °C and a heat sink at 50 °C. It should produce a net work output of 2.033 kJ. The peak cycle pressure should be 4.34 MPa and the minimum cycle pressure should be 0.12 MPa. The system should be of the non-flow type and should contain helium gas, which has a specific gas constant of 2.077 kJ/kg K and an adiabatic index of 1.667. Ideal gas behaviour should be assumed. The specification should consist of the

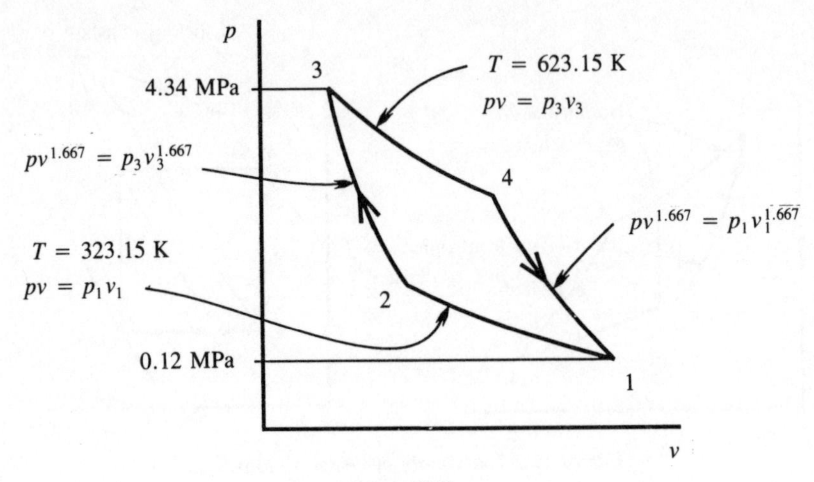

Figure 15.3

temperature, pressure and volume at each of the principal points of the cycle and the mass of gas within the system. The analysis should present the thermal efficiency of the cycle and the heat and work for each of the processes of the cycle.

SOLUTION

The cycle is shown on a $p-v$ diagram in Figure 15.3. The isothermal line that passes through points 1 and 2 is defined by the temperature of the heat sink. The isothermal line that passes through points 3 and 4 is defined by the temperature of the heat source. As the pressures at points 1 and 3 are given, these points are located. The reversible adiabatic line that passes through points 2 and 3 is fully defined by the adiabatic index and the coordinates of point 3. Similarly, the reversible adiabatic line that passes through points 4 and 1 is fully defined by the adiabatic index and the coordinates of point 1. Points 2 and 4 are located at the intersections of the isothermal lines and the reversible adiabatic lines. The required mass of gas in the system will depend on the net work output. As the net work output per unit mass can be calculated for the processes on the $p-v$ diagram, there is sufficient information to determine the mass of gas. Therefore, the cycle is fully defined by the given information.

As the system is of the non-flow type, it will be analyzed using the non-flow energy equation and expressions for the displacement work (or normal work) for a non-flow system. The ideal gas equation and ideal gas relationships will also be used.

$$T_L = 50\ [°C] + 273.15\ [K] = 323.15\ K$$

$$T_1 = T_2 = T_L = 323.15 \text{ K}$$

$$T_H = 350\ [°C] + 273.15\ [K] = 623.15 \text{ K}$$

$$T_3 = T_4 = T_H = 623.15 \text{ K}.$$

Determine the specific volumes at points 1 and 3:

$$v_1 = \frac{RT_1}{p_1} = \frac{2.077 \times 10^3\ [\text{J kg}^{-1}\ \text{K}^{-1}]\ 323.15\ [\text{K}]}{0.12 \times 10^6\ [\text{Pa}]} = 5.593 \frac{\text{m}^3}{\text{kg}}$$

$$v_3 = \frac{RT_3}{p_3} = \frac{2.077 \times 10^3\ [\text{J kg}^{-1}\ \text{K}^{-1}]\ 623.15\ [\text{K}]}{4.34 \times 10^6\ [\text{Pa}]} = 0.2982 \frac{\text{m}^3}{\text{kg}}.$$

Determine the specific volumes at points 2 and 4:

$$p_2 v_2^\gamma = p_3 v_3^\gamma \quad \text{and} \quad p_2 v_2 = p_1 v_1.$$

Eliminating p_2,

$$p_3 \left(\frac{v_3}{v_2}\right)^\gamma = p_1 \frac{v_1}{v_2}$$

$$v_2^{1-\gamma} = \frac{p_1}{p_3} v_1 v_3^{-\gamma}$$

$$v_2 = \left(\frac{p_1}{p_3} v_1 v_3^{-\gamma}\right)^{1/(1-\gamma)}$$

$$= \left(\frac{0.12\ [\text{MPa}]}{4.34\ [\text{MPa}]} 5.593 (0.2982)^{-1.667} \left[\frac{\text{m}^3}{\text{kg}}\right]^{(1-1.667)}\right)^{1/(1-1.667)}$$

$$= 0.7981 \text{ m}^3/\text{kg}.$$

Similarly,

$$v_4 = \left(\frac{p_3}{p_1} v_3 v_1^{-\gamma}\right)^{1/(1-\gamma)}$$

$$= \left(\frac{4.34\ [\text{MPa}]}{0.12\ [\text{MPa}]} 0.2982 (5.593)^{-1.667} \left[\frac{\text{m}^3}{\text{kg}}\right]^{(1-1.667)}\right)^{1/(1-1.667)}$$

$$= 2.0898 \text{ m}^3/\text{kg}.$$

Determine p_2 and p_4:

$$p_2 = p_1 \frac{v_1}{v_2} = 0.12\ [\text{MPa}]\ \frac{5.593\ [\text{m}^3\ \text{kg}^{-1}]}{0.7981\ [\text{m}^3\ \text{kg}^{-1}]} = 0.841\ \text{MPa}$$

$$p_4 = p_3 \frac{v_3}{v_4} = 4.34\ [\text{MPa}]\ \frac{0.2982\ [\text{m}^3\ \text{kg}^{-1}]}{2.0898\ [\text{m}^3\ \text{kg}^{-1}]} = 0.619\ \text{MPa}.$$

Calculate the Carnot thermal efficiency:

$$\eta_{\text{Carnot}} = \frac{T_{\text{H}} - T_{\text{L}}}{T_{\text{H}}}$$

Answer

$$= \frac{(623.15 - 323.15)\ [\text{K}]}{623.15\ [\text{K}]} = 48.14\%.$$

Calculate the heat transfer to the system:

$$Q_{\text{H,in}} = \frac{W_{\text{net,out}}}{\eta_{\text{Carnot}}} = \frac{2.033\ [\text{kJ}]}{0.4814} = 4.223\ \text{kJ}.$$

Determine the mass of gas present:
For process 3→4

$$w_{3\to4} = p_3 v_3 \ln\left(\frac{p_4}{p_3}\right)$$

$$= 4.34\ [\text{MPa}]\ 0.2982\ [\text{m}^3\ \text{kg}^{-1}]\ \ln\left(\frac{0.619\ [\text{MPa}]}{4.34\ [\text{MPa}]}\right)$$

$$= -2.52\ \text{MJ/kg}.$$

As $\Delta u_{3\to4} = 0$

$$q_{3\to4} = -w_{3\to4} = 2.52\ \text{MJ/kg}$$

$$q_{\text{H,in}} = q_{3\to4} = 2.52\ \text{MJ/kg}$$

Answer

$$m = \frac{Q_{\text{H,in}}}{q_{\text{H,in}}} = \frac{4.223\ [\text{kJ}]}{2.52 \times 10^3\ [\text{kJ kg}^{-1}]} = 1.676 \times 10^{-3}\ \text{kg}.$$

Calculate the volumes:

$$V_1 = mv_1 = 1.676 \times 10^{-3}\ [\text{kg}]\ 5.593\ [\text{m}^3\ \text{kg}^{-1}] = 9.374 \times 10^{-3}\ [\text{m}^3] = 9.374\ \text{L}$$

$$V_2 = mv_2 = 1.676 \times 10^{-3}\ [\text{kg}]\ 0.7981\ [\text{m}^3\ \text{kg}^{-1}] = 1.338 \times 10^{-3}\ [\text{m}^3] = 1.338\ \text{L}$$

$$V_3 = mv_3 = 1.676 \times 10^{-3}\ [\text{kg}]\ 0.2982\ [\text{m}^3\ \text{kg}^{-1}] = 0.500 \times 10^{-3}\ [\text{m}^3] = 0.500\ \text{L}$$

$$V_4 = mv_4 = 1.676 \times 10^{-3}\ [\text{kg}]\ 2.0897\ [\text{m}^3\ \text{kg}^{-1}] = 3.502 \times 10^{-3}\ [\text{m}^3] = 3.502\ \text{L}.$$

Answer

The properties at the principal points of the cycle are summarized in Table 15.1.

Table 15.1

Point	$\frac{T}{[\text{K}]}$	$\frac{p}{[\text{MPa}]}$	$\frac{V}{[\text{L}]}$
1	323.15	0.120	9.374
2	323.15	0.841	1.338
3	623.15	4.340	0.500
4	623.15	0.619	3.502

Calculate the work and the heat transfer for each process:

$$W_{1\to2} = p_1 V_1 \ln\left(\frac{p_2}{p_1}\right)$$

$$= 0.12 \times 10^3\ [\text{kPa}] \times 9.374 \times 10^{-3}\ [\text{m}^3] \ln\left(\frac{0.841\ [\text{MPa}]}{0.12\ [\text{MPa}]}\right) = 2.190\ \text{kJ}$$

$$Q_{1\to2} = -W_{1\to2} = -2.190\ \text{kJ}.$$

$$W_{2\to3} = \frac{p_3 V_3 - p_2 V_2}{\gamma - 1}$$

$$= \frac{(4.34 \times 0.500 - 0.841 \times 1.338) \times 10^3 \times 10^{-3}\ [\text{kPa}]\ [\text{m}^3]}{1.667 - 1} = 1.566\ \text{kJ}$$

$$Q_{2\to3} = 0.$$

$$W_{3\to4} = p_3 V_3 \ln\left(\frac{p_4}{p_3}\right)$$

$$= 4.34 \times 10^3\,[\text{kPa}] \times 0.5 \times 10^{-3}\,[\text{m}^3] \ln\left(\frac{0.619\,[\text{MPa}]}{4.34\,[\text{MPa}]}\right) = -4.226\,\text{kJ}$$

$$Q_{3\to4} = -W_{3\to4} = 4.226\ \text{kJ}.$$

$$W_{4\to1} = \frac{p_1 V_1 - p_4 V_4}{\gamma - 1}$$

$$= \frac{(0.12 \times 9.374 - 0.619 \times 3.502) \times 10^3 \times 10^{-3}\,[\text{kPa}]\,[\text{m}^3]}{1.667 - 1} = -1.564\ \text{kJ}$$

$$Q_{4\to1} = 0.$$

Answer

The heat and work quantities are summarized in Table 15.2.

Table 15.2

Process	Q [kJ]	W [kJ]
1→2	−2.190	2.190
2→3	0	1.566
3→4	4.226	−4.226
4→1	0	−1.564

15.2.1 Practical aspects of the non-flow Carnot cycle

To date, at least, no practical heat engines have been built based on the Carnot cycle for a closed system containing a gas. Various practical engines contain compression or expansion processes that are nominally adiabatic. However, isothermal compression or expansion of a gas within a closed non-flow system is very difficult to achieve, except at very low heat transfer and work rates. Such low rates would cause the engine to be very large in relation to its power output.

Another major difficulty is that the boundary of the system needs to be an excellent insulator during two of the processes, but an excellent conductor during the other two. Moreover, while the boundary serves as an excellent conductor between the heat source and the working fluid, it must provide excellent insulation from the heat sink. While it serves as an excellent conductor between the working fluid and the heat sink, it must provide excellent insulation from the heat source.

In the early days of engine development it was found that successively heating and cooling the containment of the working fluid system was not satisfactory. A lot of energy was required just to heat the containment, owing to its heat capacity. Each time the containment was heated and cooled, the energy that went to heat it was wasted. It was found much more practical to maintain the containment at a constant temperature and to move the fluid between the regions where the containment temperature was high and those where it was low. Of course, if this is done the system is no longer a non-flow system.

The Otto cycle, which is described in Chapter 17, is an example of a cycle that employs nominally adiabatic compression and nominally adiabatic expansion while the system remains closed and no flow occurs. To this extent, at least, it resembles the Carnot cycle. However, in the practical form of this cycle, as implemented in internal combustion engines, the isothermal heat transfer of the Carnot cycle is replaced by the irreversible release of fuel energy at constant volume. This avoids the difficulty of providing the heat transfer to the cycle through the containment since the fuel is part of the working fluid. The process can be roughly modelled as heat transfer at constant volume and it involves a temperature rise. The isothermal heat rejection of the Carnot cycle is replaced by an exchange of the spent charge for a fresh charge of air and fuel of the same volume. This can be modelled as roughly equivalent to heat rejection at constant volume. It involves a temperature fall. The gas exchange process avoids the difficulty of providing heat rejection through the containment.

15.3 The Carnot cycle for a two-phase working fluid

The Carnot cycle for a two-phase working fluid is illustrated in Figure 15.4. Advantage is taken of the fact that a two-phase fluid can undergo an isothermal process involving heat transfer at constant pressure when it condenses or evaporates. By restricting the cycle to the region below the saturation curve, both of its heat transfer processes occur at constant pressure. This cycle could be undergone by a non-flow system, or contained within a steady flow system. It will be described here for a steady flow system and with water substance as the working fluid.

Figure 15.5 is a schematic representation of the plant that would be necessary to implement the ideal Carnot cycle within a steady flow system. At state 1, at position 1 within the steady flow system, the water substance consists of a wet mixture

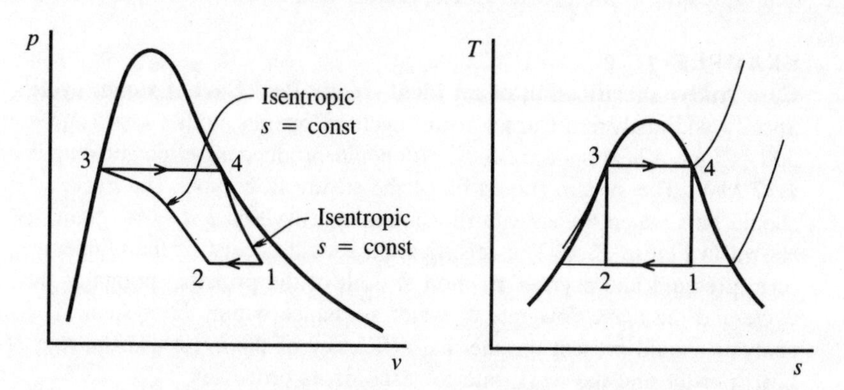

Figure 15.4 *The Carnot cycle for a two-phase working fluid.*

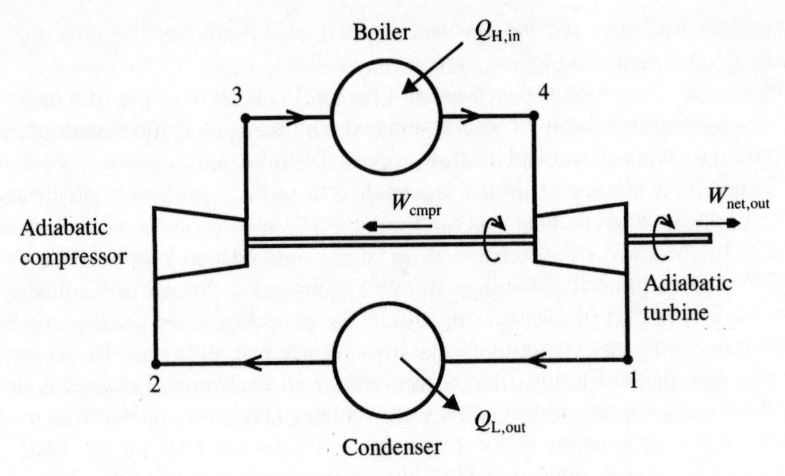

Figure 15.5 *An ideal steady flow Carnot cycle plant.*

of liquid water and vapour at the temperature of the heat sink. The dryness fraction is such that the specific entropy of the mixture is the same as that of dry saturated vapour at the temperature of the heat source. In the first process, 1→2, heat rejection occurs at constant temperature and pressure. The dryness fraction of the wet steam decreases as some of the vapour condenses. At the end of this first process the wet steam has a much lower dryness fraction than it had at state 1. The dryness fraction at state 2 is such that the wet steam has the same specific entropy as saturated liquid at the temperature of the heat source (state 3). In process 2→3 the wet steam is compressed reversibly and adiabatically (i.e. isentropically) and leaves as saturated liquid at the temperature of the heat source.

In the third process, 3→4, the saturated liquid is evaporated to the dry saturated vapour state. This occurs at constant pressure and at the temperature of the heat source. In the last process, 4→1, the steam is expanded reversibly and adiabatically (i.e. isentropically) to the pressure, temperature and dryness fraction at state 1.

EXAMPLE 15.2

Case study: specification of an ideal steady flow Carnot steam cycle

Specify and analyze a Carnot steam cycle to operate with a heat source at 300 °C and a heat sink at 20 °C. It should produce a net power output of 1.37 MW. The system should be of the steady flow type. The cycle should be situated underneath the saturation curve on a T–s diagram, as shown in Figure 15.4. The specification should consist of the temperature, pressure and dryness fraction at each of the principal points of the cycle and the mass flow rate of water substance within the system. The analysis should present the thermal efficiency of the cycle and the rate of heat transfer and the work rate for each of its processes.

SOLUTION

The isothermal line that passes through points 1 and 2 (Figure 15.4) is defined by the temperature of the heat sink. The isothermal line that passes through points 3 and 4 is defined by the temperature of the heat source. Point 3 is located on the saturated liquid line and point 4 is located on the dry saturated vapour line. Point 1 is fully defined by the temperature of the heat sink and the specific entropy at point 4. Point 2 is fully defined by the temperature of the heat sink and the specific entropy at point 3. The required mass flow rate of steam in the system will depend on the net power output. As the net power output per unit mass can be calculated from the specific enthalpy values at the principal state points, there is sufficient information to determine the mass flow rate of steam. Therefore, the cycle is fully defined by the given information.

As the system is of the steady flow type it will be analyzed using the steady flow energy equation. The thermodynamic properties at the principal state points will be found from the steam tables.

$$T_L = 20\ [°C] + 273.15\ [K] = 293.15\ K$$

$$T_1 = T_2 = T_L = 293.15\ K$$

$$T_H = 300\ [°C] + 273.15\ [K] = 573.15\ K$$

$$T_3 = T_4 = T_H = 573.15\ K.$$

From the saturation steam tables at 300 °C,

$$p_3 = p_4 = p_s = 8.593\ MPa$$

$$h_3 = h_f = 1345.1\ kJ/kg \quad \text{and} \quad s_3 = s_f = 3.255\ kJ/kg\ K$$

$$h_4 = h_g = 2751.0\ kJ/kg \quad \text{and} \quad s_4 = s_g = 5.708\ kJ/kg\ K.$$

From the saturation steam tables at 20 °C,

$$p_1 = p_2 = p_s = 0.002\ 34\ MPa$$

$$h_f = 83.9\ kJ/kg,\ h_{fg} = 2454.3\ kJ/kg,\ \text{and}\ h_g = 2538.2\ kJ/kg$$

$$s_f = 0.296\ kJ/kg\ K,\ s_{fg} = 8.372\ kJ/kg\ K,\ \text{and}\ s_g = 8.668\ kJ/kg\ K.$$

Find the dryness fraction and the specific enthalpy at state 1:

$$s_1 = s_4 = 5.708 \text{ kJ/kg K}$$

$$x_1 = \frac{s_1 - s_f}{s_{fg}} = \frac{(5.708 - 0.296)\ [\text{kJ kg}^{-1}\ \text{K}^{-1}]}{8.372\ [\text{kJ kg}^{-1}\ \text{K}^{-1}]} = 0.6464$$

$$h_1 = h_f + x h_{fg} = 83.9\ [\text{kJ kg}^{-1}] + 0.6464(2454.3)\ [\text{kJ kg}^{-1}] = 1670.4 \text{ kJ/kg.}$$

Find the dryness fraction and the specific enthalpy at state 2:

$$s_2 = s_3 = 3.255 \text{ kJ/kg K}$$

$$x_2 = \frac{s_2 - s_f}{s_{fg}} = \frac{(3.255 - 0.296)\ [\text{kJ kg}^{-1}\ \text{K}^{-1}]}{8.372\ [\text{kJ kg}^{-1}\ \text{K}^{-1}]} = 0.3534$$

$$h_2 = h_f + x_2 h_{fg} = 83.9\ [\text{kJ kg}^{-1}] + 0.3534(2454.3)\ [\text{kJ kg}^{-1}] = 951.2 \text{ kJ/kg.}$$

Answer
The properties at the principal points of the cycle are summarized in Table 15.3.

Table 15.3

Point	*t* [°C]	*p* [MPa]	*x* [%]
1	20	0.002 34	64.6
2	20	0.002 34	35.3
3	300	8.593	0
4	300	8.593	100

Apply the steady flow energy equation for each of the processes of the cycle:

$$q_{1\to2} + w_{1\to2} = h_2 - h_1.$$

But as no shear work is done on or by the fluid in a reversible constant pressure steady flow process

$$w_{1\to2} = 0.$$

Hence

$$q_{1\to2} = h_2 - h_1 = (951.2 - 1670.4) \text{ kJ/kg} = -719.2 \text{ kJ/kg.}$$

$$q_{2\to3} + w_{2\to3} = h_3 - h_2.$$

But

$$q_{2\to3} = 0.$$

Hence

$$w_{2\to3} = h_3 - h_2 = (1345.1 - 951.2) \text{ kJ/kg} = 393.9 \text{ kJ/kg}$$

$$w_{3\to4} = 0$$

$$q_{3\to4} = h_4 - h_3 = (2751.0 - 1345.1) \text{ kJ/kg} = 1405.9 \text{ kJ/kg}$$

$$q_{4\to1} = 0$$

$$w_{4\to1} = h_1 - h_4 = (1670.4 - 2751.0) \text{ kJ/kg} = -1080.6 \text{ kJ/kg}.$$

Evaluate the thermal efficiency:

$$q_{\text{H,in}} = q_{3\to4} = 1405.9 \text{ kJ/kg}$$

$$w_{\text{net,out}} = -(w_{2\to3} + w_{4\to1}) = -(393.9 - 1080.6) \text{ [kJ/kg]} = 686.7 \text{ kJ/kg}$$

$$\eta_{\text{Carnot}} = \frac{w_{\text{net,out}}}{q_{\text{H,in}}} = \frac{686.7 \text{ [kJ kg}^{-1}]}{1405.9 \text{ [kJ kg}^{-1}]} = 48.84\%.$$

As a cross-check, this can also be evaluated in terms of the temperatures of the heat source and sink:

Answer

$$\eta_{\text{Carnot}} = \frac{T_{\text{H}} - T_{\text{L}}}{T_{\text{H}}} = \frac{(573.15 - 293.15) \text{ [K]}}{573.15 \text{ [K]}} = 48.85\%.$$

Calculate the mass flow rate of steam:

Answer

$$\dot{m} = \frac{\dot{W}}{w_{\text{net,out}}} = \frac{1.37 \times 10^3 \text{ [kW]}}{686.7 \text{ [kJ kg}^{-1}]} = 1.995 \text{ kg/s}.$$

Calculate the rates of heat transfer and work:

$$\dot{Q}_{1\to2} = \dot{m}q_{1\to2} = 1.995 \text{ [kg s}^{-1}] (-719.2) \text{ [kJ kg}^{-1}] = -1.435 \times 10^3 \text{ kW}$$

$$\dot{W}_{2\to3} = \dot{m}w_{2\to3} = 1.995 \text{ [kg s}^{-1}] \, 393.9 \text{ [kJ kg}^{-1}] = 0.786 \times 10^3 \text{ kW}$$

$$\dot{Q}_{3\to4} = \dot{m}q_{3\to4} = 1.995\ [\text{kg s}^{-1}]\ 1405.9\ [\text{kJ kg}^{-1}] = 2.805 \times 10^3\ \text{kW}$$

$$\dot{W}_{4\to1} = \dot{m}w_{4\to1} = 1.995\ [\text{kg s}^{-1}]\ (-1080.3)\ [\text{kJ kg}^{-1}] = -2.156 \times 10^3\ \text{kW}.$$

Answer

The rates of heat transfer and work are summarized in Table 15.4.

Table 15.4

Process	$\dot{Q}$ [MW]	$\dot{W}$ [MW]
1→2	−1.435	0
2→3	0	0.786
3→4	2.805	0
4→1	0	−2.156

15.3.1 *Practical aspects of the steady flow Carnot cycle*

For the present, at least, no practical steam power plant operates on the Carnot cycle, owing to difficulties with its implementation. However, steam plants exist that employ three of the processes of the Carnot steam cycle: constant pressure evaporation, adiabatic expansion and constant pressure condensation. The adiabatic compression process of the Carnot cycle presents practical difficulties as compressors that would compress a two-phase mixture of liquid and vapour with a low dryness fraction to a saturated liquid exit state are not available. Also, there has been little incentive to try to develop such machines. The practical solution is to condense the vapour fully to the liquid state and then to use a pump, rather than a compressor, to increase the pressure of the fluid from the condenser pressure to the boiler pressure. The work required by the pump is much less than would be required by the adiabatic compressor of the Carnot cycle. Additional heat transfer is then required to bring the water to the saturation temperature in the boiler. The cycle is no longer one that receives all its heat transfer at a fixed temperature.

As the critical temperature of water substance is 374.15 °C, it is not possible to have isothermal heat transfer at constant pressure at temperatures higher than this. For this reason too, practical steam cycles differ from the ideal Carnot cycle. Practical steam plants are based on the Rankine cycle, which is described in Chapter 16.

15.4 Summary

The Carnot cycle has been introduced and its efficiency derived from its shape on the *T–s* diagram. The cycle has been described in some detail for a non-flow system with an ideal gas working fluid and for a steady flow system with a two-phase working fluid. The practical aspects of these cycles have been discussed. An extensive case study has been presented for each version of the cycle.

The student should be able to

- *describe*
  — the processes of the Carnot cycle, the Carnot efficiency, the Carnot cycle for a non-flow system with an ideal gas working fluid, the Carnot cycle for a steady flow system with a two-phase working fluid
- *sketch*
  — the Carnot cycle on a p–v diagram and on a T–s diagram for an ideal gas working fluid, the Carnot cycle on a p–v diagram and on a T–s diagram for a two-phase working fluid, the plant required for a steady flow Carnot cycle with a two-phase working fluid
- *derive*
  — the thermal efficiency of the Carnot cycle from its shape on the T–s diagram
- *analyze*
  — each of the processes of a Carnot cycle undergone by a non-flow system containing an ideal gas, each of the processes of a Carnot cycle within a steady flow system for a two-phase working fluid
- *discuss*
  — the practical aspects of the Carnot cycle for a non-flow system containing a gas and for a steady flow system containing a two-phase fluid.

15.5 Self-assessment questions

15.1 Suppose there exists a reciprocating engine that operates on the Carnot cycle with air as the working fluid. Let the cylinder have a maximum volume of 1 litre and a minimum volume of 0.15 litres. Let the pressure at the maximum volume position be 1 bar absolute and the temperature be 25 °C. Let the first process, isothermal heat rejection, end when the volume is 0.5 litres. Calculate the work and the heat transfer for each of the first two processes of the cycle assuming ideal gas behaviour. Also calculate the thermal efficiency of the cycle.

15.2 Suppose there exists a Carnot engine that operates using steam as the working fluid in a steady flow cycle. Let the boiler pressure be 8.0 MPa and the condenser pressure be 0.2 MPa. Let the mass flow rate of the steam be 0.843 kg/s. Let water enter the boiler as saturated liquid and steam leave the boiler as dry saturated vapour. Determine the thermal efficiency, the net power output and the rate of heat rejection in the condenser.

15.3 Saturated liquid water enters the boiler of an ideal Carnot cycle plant and dry saturated steam enters the turbine at 150 °C. If the steam leaves the turbine at 35 °C, determine the values of the dryness fraction entering the condenser and leaving the condenser.

15.4 Calculate the ratio of the volume before adiabatic compression to the volume after adiabatic compression for a non-flow Carnot cycle that operates between thermal reservoirs at 900 °C and 30 °C using air as the working fluid. Calculate the same volume ratio if argon is used.

CHAPTER 16

The Rankine steam cycle

In this chapter the Rankine steam cycle is introduced. It is made up of a number of steady flow processes. These are described, together with the basic theory for engineering calculations relating to steam power plants. Some practical aspects of the cycle are discussed.

The Rankine steam cycle was the cycle on which reciprocating steam engines operated in the golden age of steam. What is more important though, is that it is the cycle on which steam turbine plants operate today. These plants provide a very high proportion of the electricity generated in the world, whether from fossil fuels or nuclear power. All common vapour power cycles are variations on the Rankine cycle: they are based on flow processes that occur in separate devices, which are linked to form a closed circuit.

16.1 The processes of the Rankine cycle

The Rankine cycle is shown on a temperature versus specific entropy diagram in Figure 16.1. The corresponding plant components (feed pump, boiler, superheater, turbine and condenser) are shown schematically in Figure 16.2. The processes of the cycle, which are all assumed to be reversible, are as follows:

$1 \rightarrow 2$ water is pumped from the condenser pressure to the boiler pressure
$2 \rightarrow a$ heat transfer occurs to the water at constant pressure in the boiler
$a \rightarrow b$ evaporation occurs at constant pressure in the boiler
$b \rightarrow 3$ heat transfer occurs to the steam at constant pressure in the superheater
$3 \rightarrow 4$ reversible adiabatic expansion of the steam occurs in the turbine
$4 \rightarrow 1$ condensation of the exhaust steam occurs in the condenser at constant pressure.

Although a temperature rise is shown for process $1 \rightarrow 2$ in Figure 16.1, this change is very small and its size is greatly exaggerated. This is done to emphasize that this is a distinct process and to indicate that it is isentropic. If process $1 \rightarrow 2$ were shown to scale, point 2 would appear superimposed on point 1, as water liquid undergoes only a minute temperature change when its pressure is increased reversibly and adiabatically. For a perfectly incompressible fluid (which does not exist), there would be no temperature change.

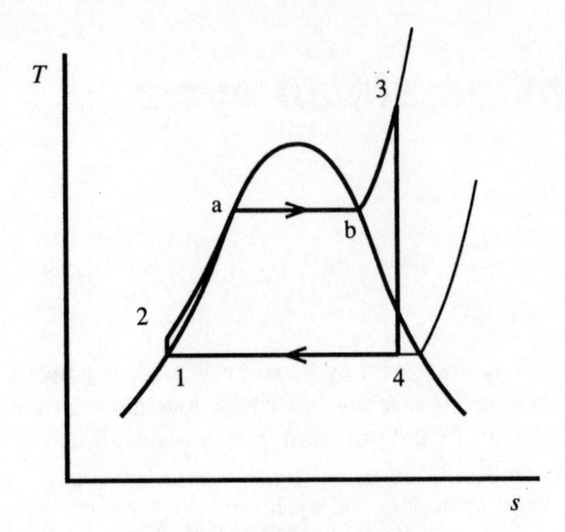

Figure 16.1 *The Rankine cycle.*

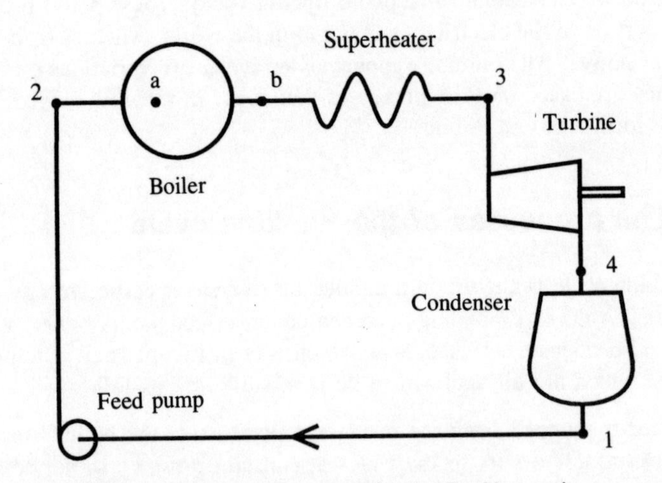

Figure 16.2 *Plant components for the Rankine cycle.*

It can be noted that most of process 3→4 is within the superheat region, but that the dryness fraction at point 4 is somewhat less than 100%. This is typically the case in practice.

16.2 Cycle calculations

The steady flow energy equation can be applied to each component of the plant to calculate the heat transfer and work for the process that occurs within it:

$$q + w + h_1 + \frac{c_1^2}{2} + gz_1 = h_2 + \frac{c_2^2}{2} + gz_2. \tag{16.1}$$

It will be assumed that the kinetic energy terms have the same value at each of the principal points in the cycle (1, 2, 3 and 4). These are the points where the fluid enters or leaves a main component. A similar assumption will be made for the potential energy terms. The steady flow energy equation for each component thus reduces to the form

$$q + w + h_1 = h_2. \tag{16.2}$$

16.2.1 Pumping

The ideal pumping process is assumed to be reversible and adiabatic. It is therefore an isentropic process. As there is no heat transfer, Equation (16.2) reduces to

$$w_{pmp} = h_2 - h_1. \tag{16.3}$$

At point 1 the specific enthalpy is that of saturated liquid at the condenser pressure. The value of h_2 cannot be read directly from standard steam tables at the boiler pressure and at $s_2 = s_1$, but can be calculated using Equation (16.6), which is explained as follows:

$$h_2 - h_1 = (u_2 - u_1) + (p_2 v_2 - p_1 v_1). \tag{16.4}$$

Also, for an incompressible fluid,

v is constant
u depends on T only and vice versa (see Chapter 6, section 6.5)
T is not changed in a reversible adiabatic pumping process (each point system experiences only a pressure change).

Therefore

$$h_2 - h_1 = (p_2 - p_1)v_1 \tag{16.5}$$

and

$$h_2 = h_1 + (p_2 - p_1)v_1. \qquad \text{(cf. 7.2, Chapter 7)} \tag{16.6}$$

Thus the ideal pumping work can conveniently be calculated from the expression

$$w_{pmp} = (p_2 - p_1)v_1. \tag{16.7}$$

The pumping power is given by the expression

$$\dot{W}_{pmp} = \dot{m}w_{pmp} = \dot{m}(p_2 - p_1)v_1. \tag{16.8}$$

16.2.2 Constant pressure heat transfer

The process in the boiler and superheater is a heating process at constant pressure. Part of the process, from a to b in Figure 16.1, is an evaporation process. As the water substance remains saturated during evaporation, the temperature remains constant for this part of the process. There is no shaft work between flow positions 2 and 3, and the steady flow energy equation reduces to

$$q_{2\to3} + h_2 = h_3. \tag{16.9}$$

Therefore

$$q_{b\&s} = q_{2\to3} = h_3 - h_2 \tag{16.10}$$

where $q_{b\&s}$ is the heat transfer to the water substance in the boiler and superheater. The specific enthalpy at point 3 can be determined from the boiler pressure (which is also the pressure in the superheater) and the temperature at the superheater exit.

16.2.3 Expansion

The expansion process in the turbine is assumed to be reversible and adiabatic. It is therefore isentropic. Equation (16.2) takes the form

$$w_{3\to4} + h_3 = h_4. \tag{16.11}$$

Hence, the work done *by* the steam on the turbine is given by

$$w_{turb,out} = -w_{3\to4} = h_3 - h_4. \tag{16.12}$$

The specific enthalpy of the steam that leaves the turbine can be found from the condenser pressure and the specific entropy. The specific entropy is the same as at point 3. The procedure for determining h_4 is therefore as follows:

- determine the specific entropy at point 3 (which is usually in the superheat region) from the temperature and pressure at that point (or, if the steam is saturated, from the temperature or pressure and the dryness fraction)
- set the specific entropy at point 4 equal to that at point 3
- determine s_f, s_{fg} and s_g and also h_f and h_{fg} at the condenser pressure
- if $s_4 < s_g$, evaluate x_4 from $x_4 = (s_4 - s_f)/s_{fg}$ and evaluate h_4 from $h_4 = h_f + x_4 h_{fg}$
- if $s_4 = s_g$, $h_4 = h_g$

- if $s_4 > s_g$, determine h_4 in the superheat region (by interpolation from the tables) from the condenser pressure and s_4.

16.2.4 Condensation

The steam that leaves the turbine is condensed at constant pressure and temperature in the condenser. There is no shaft work and Equation (16.2) takes the form

$$q_{\text{cond,out}} = -q_{4\to1} = h_4 - h_1 \qquad (16.13)$$

where $q_{\text{cond,out}}$ is the heat *rejection* from the steam in the condenser.

16.3 Cycle thermal efficiency

The thermal efficiency of the cycle is given by the following expression:

$$\eta_{\text{th}} = \frac{\text{net work output}}{\text{heat input}} = \frac{w_{\text{turb,out}} - w_{\text{pmp}}}{q_{\text{b\&s}}} \qquad (16.14)$$

$$= \frac{(h_3 - h_4) - (h_2 - h_1)}{h_3 - h_2} \qquad (16.15)$$

$$= \frac{(h_3 - h_4) - (p_2 - p_1)v_1}{h_3 - h_1 - (p_2 - p_1)v_1}. \qquad (16.16)$$

EXAMPLE 16.1

Case study: a steam power plant

An ideal Rankine cycle plant operates with a boiler pressure of 5 MPa and a condenser pressure of 0.015 MPa. Steam leaves the superheater at 400 °C. Work is done on the turbine by the steam at the rate of 500 kW. Determine the following:

(a) the dryness fraction of the exhaust steam
(b) the steam mass flow rate
(c) the ideal pumping power
(d) the cycle efficiency
(e) the rate of heat rejection in the condenser
(f) the minimum diameter of the pipe between the turbine and the condenser if the steam velocity is not to exceed 10 m/s.

SOLUTION

The cycle is shown on a T–s diagram in Figure 16.1. At point 3, p = 5 MPa, t = 400 °C. From the superheat tables, h_3 = 3198.3 kJ/kg and s_3 = 6.651 kJ/kg K. At the condenser pressure of 0.015 MPa, from the saturation tables

$$h_1 = h_f = 226.0 \text{ kJ/kg} \quad h_{fg} = 2373.2 \text{ kJ/kg}$$

$$s_f = 0.755 \text{ kJ/kg K} \quad s_{fg} = 7.254 \text{ kJ/kg K} \quad s_g = 8.009 \text{ kJ/kg K}$$

$$v_1 = v_f = 0.001\,014 \text{ m}^3\text{/kg} \quad v_{fg} = 10.021\,815 \text{ m}^3\text{/kg}.$$

To find the dryness fraction at point 4:

$$s_4 = s_3 = 6.651 \text{ kJ/kg K}.$$

Since $s_4 < s_g$, the steam at point 4 is saturated.

Answer (a)

$$x_4 = \frac{s_4 - s_f}{s_{fg}} = \frac{6.651 - 0.755}{7.254} = 0.8128.$$

Therefore, the dryness fraction of the exhaust steam is 81.3%.

$$h_4 = h_f + x_4 h_{fg}$$

$$= (226.0 + 0.8128 \times 2373.2) \text{ kJ/kg}$$

$$= 2154.9 \text{ kJ/kg}.$$

The turbine power output is given by

$$\dot{W}_{turb,out} = \dot{m}(h_3 - h_4).$$

Hence

$$\dot{m} = \frac{\dot{W}_{turb,out}}{(h_3 - h_4)}$$

$$= \frac{500 \text{ [kW]}}{(3198.3 - 2154.9) \text{ [kJ kg}^{-1}\text{]}}$$

Answer (b)

$$= 0.479 \text{ kg/s}.$$

The pump work input per unit mass is

$$w_{pmp} = (p_2 - p_1)v_1.$$

$$v_1 = v_f = 0.001\,014 \text{ m}^3/\text{kg}.$$

Hence

$$w_{pmp} = (5.0 - 0.015) \times 10^3 \times 0.001\,014 \text{ [kNm}^{-2}\text{] [m}^3\text{ kg}^{-1}\text{]}$$

$$= 5.05 \text{ kJ/kg}.$$

The pumping power input is given by

Answer (c)

$$\dot{W}_{pmp} = \dot{m}w_{pmp} = 0.479 \times 5.05 \text{ [kg s}^{-1}\text{] [kJ kg}^{-1}\text{]}$$

$$= 2.42 \text{ kW}.$$

$$E_{th} = \frac{(h_3 - h_4) - w_{pmp}}{(h_3 - h_1) - w_{pmp}}$$

$$= \frac{(3198.3 - 2154.9 - 5.05) \text{ [kJ kg}^{-1}\text{]}}{(3198.3 - 226.0 - 5.05) \text{ [kJ kg}^{-1}\text{]}}$$

Answer (d)

$$= 0.350.$$

Therefore, the thermal efficiency is 35.0%.

The rate of heat rejection in the condenser is given by

$$\dot{Q}_{cond,out} = \dot{m}(h_4 - h_1)$$

$$= 0.479(2154.9 - 226.0) \text{ [kg s}^{-1}\text{] [kJ kg}^{-1}\text{]}$$

Answer (e)

$$= 924 \text{ kW}.$$

The specific volume of the exhaust steam at point 4 is given by

$$v_4 = v_f + x_4 v_{fg} = 0.001\,014 + 0.8128 \times 10.021\,815 \text{ m}^3/\text{kg}$$

$$= 8.147 \text{ m}^3/\text{kg}.$$

The mass flow rate is related to the pipe diameter D_4 by

$$\dot{m} = \frac{A_4 c_4}{v_4} = \frac{\pi D_4^2 c_4}{4 v_4}.$$

Hence, if $c_4 = 10$ m/s,

Answer (f)

$$D_4 = \sqrt{\frac{4 v_4 \dot{m}}{\pi c_4}} = \sqrt{\frac{4 \times 8.147 \times 0.479 \text{ [m}^3\text{ kg}^{-1}\text{] [kg s}^{-1}\text{]}}{\pi \times 10 \text{ [m s}^{-1}\text{]}}} = 0.705 \text{ m}.$$

As a smaller diameter would increase the velocity of the exhaust steam, this is the minimum required diameter.

16.4 Practical aspects of the Rankine steam cycle

High temperatures are desirable for the heat transfer to the working fluid in order to maximize the thermal efficiency of the cycle. If the dryness fraction of the steam that leaves the turbine is too low, particles of liquid in the wet steam can cause erosion of the turbine blades. This is another reason why a high superheat temperature is desirable. The maximum cycle temperature (t_3) is limited, at present, to about 650 °C in practical steam cycles. This is due to the unavailability of superheater tube materials that could withstand higher temperatures while under stress (due to the pressure of the steam).

The condensing temperature must always be above that of the heat sink, which would typically be environmental air or sea water. Thus, the condensing temperature might be in the range 20 °C to 40 °C. The corresponding saturation pressure would be well below atmospheric; that is, the condenser would operate under vacuum.

The efficiency of an actual plant would be somewhat lower than that of the ideal plant described in this chapter for the same boiler and condenser pressures and the same maximum cycle temperature. The difference in performance would mainly be due to turbine inefficiency, caused by fluid friction and mechanical friction.

The practical form of the Rankine cycle is a true heat engine. The steady flow system that contains the steam receives part of the input heat transfer at constant temperature and part over a temperature range. It rejects heat at constant temperature to a heat sink.

A typical source of the heat transfer to a Rankine steam cycle is the steady flow combustion of a fuel in air. A combustible mixture of air and fuel is in an unstable equilibrium state. When this is ignited, combustion occurs. There is a major change in its state, including a change in its chemical composition. This irreversible process involves chemical reaction and the release of chemical energy. If this change occurred adiabatically and at constant pressure, the temperature of the combustion products would reach a very high value, known as the *adiabatic flame temperature*. In a fuel-fired steam power plant, the combustion products can be regarded as being cooled from the adiabatic flame temperature to their final exit temperature as they provide heat transfer to the steam. The temperature at which heat is provided from the heat source thus varies from the adiabatic flame temperature to the final exhaust temperature of the combustion products. The temperature at which heat is received by the steam varies from t_2 to t_3, as shown in Figure 16.1. The heat transfer from the heat source to the steam over any small area of the heat transfer surface typically involves a large temperature difference and is thus irreversible.

16.5 Practical tips

- For each of the main plant components for an ideal Rankine steam cycle, the steady flow energy equation can be used to express either the heat transfer or the work as a function of the enthalpy change from inlet to outlet of that component.
- The ideal feed pump work input per unit mass, which is also the difference in specific enthalpy across the ideal feed pump, is given by the term $(p_2 - p_1)v_1$.
- The specific enthalpy after the ideal expansion process can be found by first solving for the dryness fraction using $s_4 = s_3$ and the relationship $x_4 = (s_4 - s_f)/s_{fg}$.

16.6 Summary

The Rankine steam cycle has been explained. The steady flow processes that comprise it have been described and expressions derived for the work or heat transfer in each. An expression for the efficiency of the cycle has been presented. A case study has shown how a typical cycle analysis is undertaken. Practical aspects of the cycle have been discussed.

The student should be able to

- *describe*
  — the processes of the Rankine cycle
- *sketch*
  — the Rankine cycle on a T–s diagram, the plant diagram for a Rankine steam cycle
- *explain*
  — the expression for the pumping work in terms of the pressure difference and the specific volume of liquid water, the adiabatic flame temperature
- *analyze*
  — each of the processes of a Rankine cycle.

16.7 Self-assessment questions

16.1 An ideal Rankine steam cycle plant operates with a boiler pressure of 8 MPa absolute and a condenser pressure of 0.010 MPa absolute. The steam enters the turbine at 550 °C. If the mass flow rate of the steam is 3.34 kg/s, at what rate is heat rejected by the steam in the condenser?

16.2 A steam power plant operates on a Rankine cycle without superheat. The steam is in the dry saturated vapour state as it enters the turbine. Determine the ideal cycle efficiency if the boiler pressure is 1 MPa and the condensing pressure is 0.002 MPa.

16.3 The boiler pressure in a Rankine cycle steam plant is 100 bar. The condenser pressure is 3 bar. The superheat temperature at entry to the turbine is 500 °C. The turbine power is 100 MW. Find the volume flow rate of steam at entry to the turbine and at exit from the turbine. Assume reversible adiabatic expansion.

CHAPTER 17

The air standard Otto cycle

An air standard cycle is a simplified model that permits the ideal characteristics of a particular type of engine to be studied. The air standard Otto cycle is an analogue, based on ideal reversible processes, of the cycle of a spark ignition engine.

In this chapter the air standard Otto cycle and the simplifying assumptions that underlie it are described. The calculations for the heat transfer, work and thermal efficiency are presented. The air standard Otto cycle serves as a practical example of how the non-flow energy equation and the expression for polytropic work are applied for a closed system containing an ideal gas.

Although the cycle is an idealized one, the same basic principles could be extended to include non-ideal effects such as friction and irreversible heat transfer. Many of the features of real internal combustion engines such as combustion, gas exchange and flow throttling at the inlet and exhaust valves could be included. To obtain close agreement with the characteristics of an actual engine, the necessary detail in the model would be great and the calculations would be extensive. A computer would inevitably be used. This is done by specialists in this area and computer simulation models, which perform detailed cycle analyses, are an important aspect of engine design and development today.

17.1 Assumptions

In an air standard Otto cycle the working fluid is assumed to be air. This represents a significant simplification. In an Otto cycle internal combustion engine the working fluid changes from being a mixture of air and fuel to being a mixture of combustion products consisting mainly of nitrogen, water vapour and carbon dioxide. The following are the main assumptions that are made for the air standard Otto cycle:

1. The working fluid is air.
2. The air is an ideal gas and has constant specific heat values.
3. The system that undergoes the cycle is closed.
4. The processes of the cycle are reversible.

In a spark ignition internal combustion engine, combustion occurs over a relatively short period while the piston is near the top-dead-centre position. There is an irreversible release of chemical energy (internal energy associated with the bonds between the atoms that form molecules) and the temperature of the working fluid increases considerably. In the air standard Otto cycle this is modelled as heat transfer to the air at constant volume.

The fuel can be burnt only once in a real engine and the resulting combustion products are depleted of oxygen. In order for the cycle to be repeated, the spent gases must be exchanged for a new charge of air and fuel. The exhaust products leave the engine at quite a high temperature and the fresh charge enters at ambient temperature. This needs to occur when the piston is near the bottom-dead-centre position, as at this position the maximum energy will have been extracted as work, by expansion, from the hot gases. This exchange of a hot gas mixture for a cool gas mixture is modelled as heat rejection from the air at constant volume in the air standard Otto cycle.

The air standard Otto cycle has two processes in common with the Carnot cycle: reversible adiabatic compression and reversible adiabatic expansion. In these processes the temperature of the working fluid rises and falls respectively while no heat transfer occurs.

17.2 The air standard Otto cycle

The air standard Otto cycle, Figure 17.1, is made up of four reversible processes as follows:

1→2 adiabatic compression (work is done on the air), pV^γ = const
2→3 heat transfer to the air at constant volume
3→4 adiabatic expansion (work is done by the air), pV^γ = const
4→1 heat rejection at constant volume.

17.3 Cycle analysis

For the reversible adiabatic compression process,

$$p_1 V_1^\gamma = p_2 V_2^\gamma \tag{17.1}$$

$$\frac{p_2}{p_1} = \left(\frac{V_1}{V_2}\right)^\gamma. \tag{17.2}$$

From the ideal gas equation,

$$\frac{p_1 V_1}{T_1} = \frac{p_2 V_2}{T_2} \tag{17.3}$$

$$\frac{T_2}{T_1} = \frac{p_2 V_2}{p_1 V_1}. \tag{17.4}$$

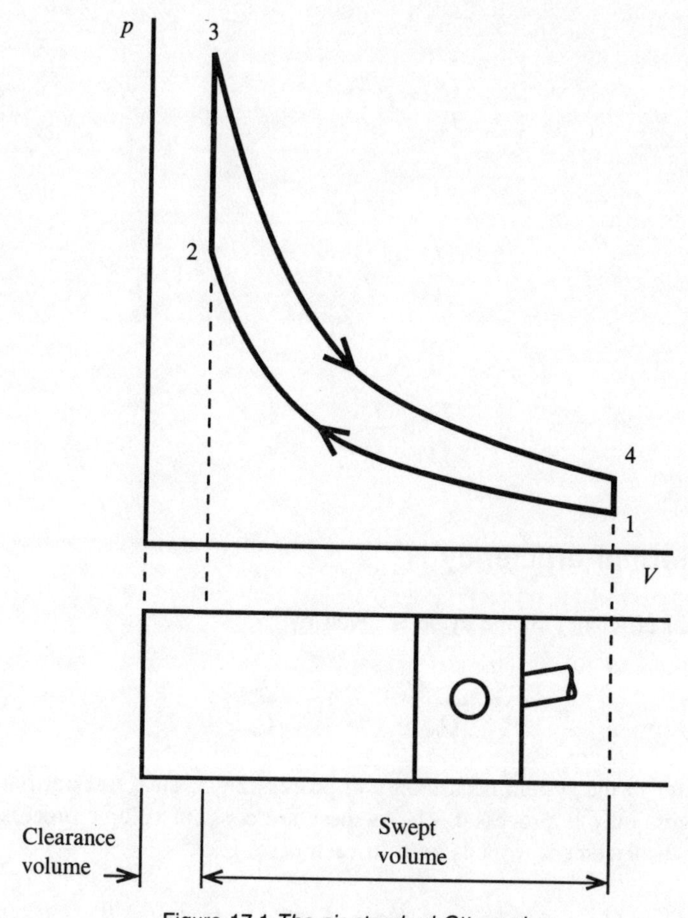

Figure 17.1 *The air standard Otto cycle.*

Combining Equations (17.2) and (17.4)

$$\frac{T_2}{T_1} = \left(\frac{V_1}{V_2}\right)^{\gamma-1}. \qquad \text{(cf. 8.24, Chapter 8)} \qquad (17.5)$$

The *compression ratio* is defined as follows:

$$r_v = \frac{V_1}{V_2} = \frac{\text{clearance volume + swept volume}}{\text{clearance volume}}. \qquad (17.6)$$

Therefore, from Equation (17.5)

$$\frac{T_2}{T_1} = r_v^{\gamma-1}. \qquad (17.7)$$

Similarly,

$$\frac{T_3}{T_4} = r_v^{\gamma-1}. \tag{17.8}$$

Therefore

$$\frac{T_2}{T_1} = \frac{T_3}{T_4} \tag{17.9}$$

and

$$\frac{T_3}{T_2} = \frac{T_4}{T_1}. \tag{17.10}$$

17.4 Thermal efficiency

The thermal efficiency of the cycle is given by

$$\eta_{th} = \frac{Q_{in} - Q_{out}}{Q_{in}} = 1 - \frac{Q_{out}}{Q_{in}}. \tag{17.11}$$

Heat transfer to the system occurs only in process $2 \rightarrow 3$. Heat transfer out of the system occurs only in process $4 \rightarrow 1$. As these are constant volume processes, the normal or displacement work is zero in each case.

$$Q_{in} = Q_{2\rightarrow 3} = \Delta U_{2\rightarrow 3} - W_{2\rightarrow 3} = mc_v(T_3 - T_2) - 0 \tag{17.12}$$

$$Q_{out} = -Q_{4\rightarrow 1} = -\Delta U_{4\rightarrow 1} + W_{4\rightarrow 1} = mc_v(T_4 - T_1) + 0. \tag{17.13}$$

Therefore

$$\eta_{th} = 1 - \frac{mc_v(T_4 - T_1)}{mc_v(T_3 - T_2)} \tag{17.14}$$

$$= 1 - \frac{T_1[(T_4/T_1) - 1]}{T_2[(T_3/T_2) - 1]}. \tag{17.15}$$

But $T_4/T_1 = T_3/T_2$ from Equation (17.10). Therefore,

$$\eta_{th} = 1 - \frac{T_1}{T_2}. \tag{17.16}$$

Also, combining Equation (17.16) with Equation (17.7),

$$\eta_{\text{th}} = 1 - \frac{1}{r_v^{\gamma-1}}. \tag{17.17}$$

EXAMPLE 17.1

Case study: idealized cycle analysis of a 1.6 litre engine

An ideal air standard Otto engine has a maximum cylinder volume of 1.6 litres, a compression ratio of 10 and a peak cycle temperature of 2200 °C. The minimum cycle temperature and pressure are 30 °C and 1 bar absolute respectively.

(a) Determine the thermal efficiency, the heat transfer to the air per cycle and the net work output per cycle.
(b) What is the average power output if the engine operates at the rate of 3000 cycles per minute?
(c) Calculate the maximum cycle pressure.

SOLUTION

Figure 17.1 illustrates the cycle. The thermal efficiency is given by

$$\eta_{\text{th}} = 1 - \frac{1}{r_v^{\gamma-1}}$$

Answer (a)

$$= 1 - \frac{1}{10^{0.4}} = 0.6019 = 60.2\%.$$

The mass of air in the cylinder is determined from the ideal gas equation, in which the temperature must be in absolute units:

$$T_1 = 30\ [°\text{C}] + 273.15\ [\text{K}] = 303.15\ \text{K}$$

$$m = \frac{p_1 V_1}{RT_1} = \frac{10^5 \times 1.6 \times 10^{-3}\ [\text{N m}^{-2}]\ [\text{m}^3]}{287 \times 303.15\ [\text{J kg}^{-1}\ \text{K}^{-1}]\ [\text{K}]}$$

$$= 1.839 \times 10^{-3}\ \text{kg}.$$

The temperature at the end of the compression process is found from the temperature at the start of the process and the compression ratio, as follows:

$$T_2 = r_v^{\gamma-1} T_1$$

$$= (10)^{0.4} 303.15\ \text{K} = 761.48\ \text{K}.$$

Also

$$T_3 = 2200\ [°C] + 273.15\ [K] = 2473.15\ K.$$

The heat transfer to the air in process 2→3 is given by

$$Q_{in} = Q_{2\rightarrow 3} = \Delta U_{2\rightarrow 3} = mc_v(T_3 - T_2)$$

$$= 1.839 \times 10^{-3} \times 718 \times (2473.15 - 761.48)\ [kg]\ [J\ kg^{-1}\ K^{-1}]\ [K]$$

Answer (a)

$$= 2260.1\ J.$$

The net work output per cycle is given by

$$W_{net,out} = Q_{2\rightarrow 3}E_{th}$$

$$= 2260.1(0.6019)\ J$$

Answer (a)

$$= 1360.4\ J.$$

At 3000 cycles per minute, the average power output is given by

$$\dot{W}_{net,out} = \frac{3000}{60}1360.4\frac{[cycles\ min^{-1}]}{[s\ min^{-1}]}\ [J]$$

Answer (b)

$$= 68\,020\ W = 68.02\ kW.$$

From the ideal gas equation

$$\frac{p_1V_1}{T_1} = \frac{p_3V_3}{T_3}.$$

Hence

$$p_3 = p_1\frac{V_1}{V_3}\frac{T_3}{T_1}$$

Answer (c)

$$= 0.1(10)\frac{2473.15}{303.15}\ MN/m^2 = 8.158\ MN/m^2.$$

17.5 Practical aspects of the air standard Otto cycle

As γ is constant, the efficiency depends only on the compression ratio. For spark ignition engines, r_v is usually in the range 8 to 10. It is limited by the occurrence of *pre-ignition*: if the air and fuel mixture were compressed any more than this the temperature rise would cause it to self-ignite before the piston reached the top-dead-centre position.

In spark ignition engines the peak gas temperature is typically about 2200 °C. Metal components are cooled and remain at a relatively low temperature. They are not in thermal equilibrium with the gas.

The ideal air standard Otto cycle is a true engine cycle. It accepts heat transfer over a temperature range with a high mean temperature and rejects heat over a temperature range with a lower mean temperature. The thermal efficiency of a spark ignition engine is related to the thermal efficiency of the corresponding air standard Otto cycle. In both cases the thermal efficiency is dependent on the compression ratio. The efficiency of the actual internal combustion engine would be lower than that of the ideal air standard Otto heat engine for the following reasons:

- In a practical spark ignition engine much work input is required to exchange the exhaust products for the fresh charge of fuel and air.
- The working fluid of the practical engine is not in mechanical, thermal or chemical equilibrium during the cycle. From Carnot's principle, a reduction in efficiency would be expected.
- There is significant heat transfer from the working fluid to the containment in an actual spark ignition engine. As this represents an increase in the heat rejection from the cycle, the net work output is reduced.
- In the practical engine the combustion process takes place over a finite period of time while the piston is moving. The release of chemical energy, which corresponds to the acceptance of heat transfer in the ideal air standard cycle, is therefore not at constant volume. It can be shown that this tends to reduce the thermal efficiency.

17.6 Practical tips

- The non-flow energy equation can be applied to determine either the heat transfer or the work for each of the processes of the air standard Otto cycle. None of the processes involves both heat and work.
- Changes in internal energy can be calculated from the temperature changes and the specific heat *at constant volume*.
- It may be useful to remember the relationship between the temperature ratio and the volume ratio (the compression ratio) for the compression and expansion processes of the air standard Otto cycle. However, you should make sure that you can derive this relationship quickly from the ideal gas equation and the polytropic relationship.
- Take care to use absolute temperatures and pressures where these are necessary.

17.7 Summary

The air standard Otto cycle and the associated engineering calculations have been described and applied in a case study. This idealized cycle has provided a thermodynamic insight into the cycle of the spark ignition internal combustion engine. The importance of the compression ratio in determining the thermal efficiency of the cycle has been noted.

The student should be able to

- *describe*
  — the processes of the air standard Otto cycle
- *sketch*
  — the air standard Otto cycle on a $p-V$ diagram
- *explain*
  — the relationship of the air standard Otto cycle to a reciprocating, internal combustion, spark ignition engine
  — pre-ignition
- *define*
  — the compression ratio
- *derive*
  — expressions for the thermal efficiency of the air standard Otto cycle in terms of the temperatures before and after compression and in terms of the compression ratio
- *analyze*
  — each of the processes of an air standard Otto cycle.

17.8 Self-assessment questions

17.1 Determine the temperature after compression and the peak cycle temperature in an ideal air standard Otto cycle if the temperature is 14 °C at the start of the compression stroke, the compression ratio is 9 and the heat transfer to the air is 1200 kJ/kg.

17.2 For an ideal air standard Otto cycle the maximum volume is 500 cubic centimetres and the compression ratio is 8. The maximum and minimum cycle temperatures are 1900 °C and 90 °C respectively. The minimum cycle pressure is 0.9 bar absolute. Determine
(a) the mass of air present
(b) the heat transfer to the system per cycle
(c) the thermal efficiency
(d) the net work per cycle.

Appendix A

Steam tables

This appendix consists of three tables. Table A1 provides the thermodynamic properties of saturated liquid water and dry saturated steam. Table A2 provides the thermodynamic properties of superheated steam at pressures up to the critical value. Table A3 provides the properties of water substance at supercritical pressures.

The tables were prepared using specially written computer programs that called FORTRAN property subroutines published by the American Society of Mechanical Engineers. These subroutines were obtained from the ASME on a diskette that came with the ASME Steam Tables, sixth edition, 1993. No warranty of accuracy or fitness for any particular purpose is made for the data.

Table A1 *Saturation table for water substance*

p_s [MPa]	t_s [°C]	u_f [kJ/kg]	u_{fg} [kJ/kg]	u_g [kJ/kg]	h_f [kJ/kg]	h_{fg} [kJ/kg]	h_g [kJ/kg]
0.000 611	**0.00**	−0.04	2375.6	2375.5	−0.04	2501.6	2501.6
0.000 611	**0.01**	0.00	2375.6	2375.6	0.00	2501.6	2501.6
0.001 00	7.0	29.3	2355.8	2385.2	29.3	2485.0	2514.4
0.001 23	**10.0**	42.0	2347.3	2389.3	42.0	2477.9	2519.9
0.002 00	17.5	73.5	2326.2	2399.6	73.5	2460.2	2533.7
0.002 34	**20.0**	83.9	2319.2	2403.0	83.9	2454.3	2538.2
0.003 00	24.1	101.0	2307.7	2408.7	101.0	2444.7	2545.7
0.004 00	29.0	121.4	2293.9	2415.3	121.4	2433.1	2554.5
0.004 24	**30.0**	125.7	2291.0	2416.7	125.7	2430.7	2556.4
0.005 00	32.9	137.8	2282.9	2420.6	137.8	2423.8	2561.6
0.006 00	36.2	151.5	2273.6	2425.1	151.5	2416.0	2567.5
0.007 00	39.0	163.4	2265.5	2428.9	163.4	2409.2	2572.6
0.007 38	**40.0**	167.4	2262.8	2430.2	167.5	2406.9	2574.4
0.008 00	41.5	173.9	2258.4	2432.3	173.9	2403.3	2577.1
0.009 00	43.8	183.3	2252.0	2435.3	183.3	2397.9	2581.1
0.0100	45.8	191.8	2246.2	2438.0	191.8	2393.0	2584.8
0.0123	**50.0**	209.2	2234.3	2443.6	209.3	2382.9	2592.2
0.0150	54.0	226.0	2222.9	2448.9	226.0	2373.2	2599.2
0.0199	**60.0**	251.1	2205.7	2456.8	251.1	2358.6	2609.7
0.0200	60.1	251.4	2205.4	2456.9	251.5	2358.4	2609.9
0.0300	69.1	289.3	2179.3	2468.6	289.3	2336.1	2625.4
0.0312	**70.0**	292.9	2176.7	2469.7	293.0	2334.0	2626.9
0.0400	75.9	317.6	2159.5	2477.1	317.7	2319.2	2636.9
0.0474	**80.0**	334.9	2147.4	2482.3	334.9	2308.8	2643.8
0.0500	81.4	340.5	2143.5	2484.0	340.6	2305.4	2646.0
0.0600	86.0	359.9	2129.8	2489.7	349.9	2293.6	2653.6
0.0700	90.0	376.7	2117.8	2494.5	376.8	2283.3	2660.1
0.0701	**90.0**	376.9	2117.7	2494.6	376.9	2283.2	2660.1
0.0800	93.5	391.6	2107.2	2498.8	391.7	2274.1	2665.8
0.0900	96.7	405.1	2097.5	2502.6	405.2	2265.7	2670.9
0.100	99.6	417.4	2088.7	2506.1	417.5	2257.9	2675.4
0.101	**100.0**	419.0	2087.5	2506.5	419.1	2256.9	2676.0
0.143	**110.0**	461.2	2056.8	2518.0	461.3	2230.0	2691.3
0.150	111.4	467.0	2052.5	2519.5	467.1	2226.2	2693.4
0.199	**120.0**	503.5	2025.5	2529.0	503.7	2202.2	2706.0
0.200	120.2	504.5	2024.7	2529.2	504.7	2201.6	2706.3
0.250	127.4	535.1	2001.7	2536.8	535.3	2181.1	2716.4
0.270	**130.0**	546.0	1993.4	2539.4	546.3	2173.6	2719.9
0.300	133.5	561.1	1981.9	2543.0	561.4	2163.2	2724.7
0.350	138.9	583.9	1964.3	2548.2	584.3	2147.4	2731.6

Table A1 *(Cont'd)*

p_s [MPa]	t_s [°C]	s_f [kJ/kg K]	s_{fg} [kJ/kg K]	s_g [kJ/kg K]	v_f [m^3/kg]	v_{fg} [m^3/kg]	v_g [m^3/kg]
0.000 611	**0.00**	0.000	9.158	9.158	0.001 000	206.304 154	206.305
0.000 611	**0.01**	0.000	9.158	9.158	0.001 000	206.161 874	206.163
0.001 00	7.0	0.106	8.871	8.977	0.001 000	129.208 040	129.209
0.001 23	**10.0**	0.151	8.751	8.902	0.001 000	106.428 772	106.430
0.002 00	17.5	0.261	8.464	8.725	0.001 001	67.005 157	67.006
0.002 34	**20.0**	0.296	8.372	8.668	0.001 002	57.837 307	57.838
0.003 00	24.1	0.354	8.224	8.579	0.001 003	45.666 298	45.667
0.004 00	29.0	0.423	8.053	8.476	0.001 004	34.801 214	34.802
0.004 24	**30.0**	0.437	8.018	8.455	0.001 004	32.927 899	32.929
0.005 00	32.9	0.476	7.920	8.396	0.001 005	28.193 351	28.194
0.006 00	36.2	0.521	7.810	8.331	0.001 006	23.740 033	23.741
0.007 00	39.0	0.559	7.718	8.277	0.001 007	20.530 034	20.531
0.007 38	**40.0**	0.572	7.686	8.258	0.001 008	19.545 098	19.546
0.008 00	41.5	0.593	7.637	8.230	0.001 008	18.103 580	18.105
0.009 00	43.8	0.622	7.566	8.188	0.001 009	16.203 265	16.204
0.0100	45.8	0.649	7.502	8.151	0.001 010	14.673 601	14.675
0.0123	**50.0**	0.704	7.374	8.078	0.001 012	12.044 694	12.046
0.0150	54.0	0.755	7.254	8.009	0.001 014	10.021 815	10.023
0.0199	**60.0**	0.831	7.080	7.911	0.001 017	7.677 521	7.679
0.0200	60.1	0.832	7.077	7.909	0.001 017	7.648 750	7.650
0.0300	69.1	0.944	6.825	7.770	0.001 022	5.228 280	5.229
0.0312	**70.0**	0.955	6.802	7.757	0.001 023	5.045 243	5.046
0.0400	75.9	1.026	6.645	7.671	0.001 027	3.992 397	3.993
0.0474	**80.0**	1.075	6.538	7.613	0.001 029	3.408 057	3.409
0.0500	81.4	1.091	6.504	7.595	0.001 030	3.239 192	3.240
0.0600	86.0	1.145	6.387	7.533	0.001 033	2.730 722	2.732
0.0700	90.0	1.192	6.288	7.480	0.001 036	2.363 696	2.365
0.0701	**90.0**	1.193	6.287	7.480	0.001 036	2.360 260	2.361
0.0800	93.5	1.233	6.202	7.435	0.001 039	2.085 924	2.087
0.0900	96.7	1.270	6.126	7.395	0.001 041	1.868 151	1.869
0.100	99.6	1.303	6.057	7.360	0.001 043	1.692 687	1.694
0.101	**100.0**	1.307	6.049	7.355	0.001 044	1.671 953	1.673
0.143	**110.0**	1.419	5.820	7.239	0.001 052	1.208 885	1.210
0.150	111.4	1.434	5.790	7.223	0.001 053	1.157 984	1.159
0.199	**120.0**	1.528	5.602	7.129	0.001 061	0.890 463	0.8915
0.200	120.2	1.530	5.597	7.127	0.001 061	0.884 380	0.8854
0.250	127.4	1.607	5.445	7.052	0.001 068	0.717 372	0.7184
0.270	**130.0**	1.634	5.392	7.026	0.001 070	0.667 066	0.6681
0.300	133.5	1.672	5.319	6.991	0.001 074	0.604 489	0.6056
0.350	138.9	1.727	5.212	6.939	0.001 079	0.522 924	0.5240

Table A1 *(Cont'd)*

p_s [MPa]	t_s [°C]	u_f [kJ/kg]	u_{fg} [kJ/kg]	u_g [kJ/kg]	h_f [kJ/kg]	h_{fg} [kJ/kg]	h_g [kJ/kg]
0.361	**140.0**	588.7	1960.6	2549.3	589.1	2144.0	2733.1
0.400	143.6	604.2	1948.5	2552.7	604.7	2133.0	2737.6
0.450	147.9	622.7	1934.0	2556.7	623.2	2119.7	2742.9
0.476	**150.0**	631.6	1926.9	2558.6	632.2	2113.2	2745.4
0.500	151.8	639.6	1920.6	2560.2	640.1	2107.4	2747.5
0.600	158.8	669.8	1896.4	2566.2	670.4	2085.0	2755.5
0.618	**160.0**	674.8	1892.3	2567.1	675.5	2081.3	2756.7
0.700	165.0	696.3	1874.8	2571.1	697.1	2064.9	2762.0
0.792	**170.0**	718.2	1856.7	2575.0	719.1	2047.9	2767.1
0.800	170.4	720.0	1855.2	2575.3	720.9	2046.5	2767.5
0.900	175.4	741.6	1837.2	2578.8	742.6	2029.5	2772.1
1.000	179.9	761.5	1820.4	2581.9	762.6	2013.6	2776.2
1.003	**180.0**	762.0	1820.0	2582.0	763.1	2013.2	2776.3
1.255	**190.0**	806.1	1782.0	2588.1	807.5	1976.7	2784.3
1.500	198.3	842.9	1749.5	2592.4	844.7	1945.2	2789.9
1.555	**200.0**	850.6	1742.7	2593.2	852.4	1938.6	2790.9
1.908	**210.0**	895.5	1701.8	2597.3	897.7	1898.5	2796.2
2.000	212.4	906.2	1691.9	2598.2	908.6	1888.6	2797.2
2.320	**220.0**	940.9	1659.4	2600.3	943.7	1856.3	2799.9
2.500	223.9	959.0	1642.2	2601.2	962.0	1839.0	2800.9
2.798	**230.0**	986.9	1615.2	2602.1	990.3	1811.7	2802.0
3.000	233.8	1004.7	1597.7	2602.4	1008.4	1793.9	2802.3
3.348	**240.0**	1033.5	1569.0	2602.5	1037.6	1764.6	2802.2
3.500	242.5	1045.4	1556.9	2602.4	1049.8	1752.2	2802.0
3.978	**250.0**	1080.8	1520.6	2601.4	1085.8	1714.7	2800.4
4.000	250.3	1082.4	1519.0	2601.3	1087.4	1712.9	2800.3
4.500	257.4	1116.4	1483.1	2599.5	1122.1	1675.6	2797.7
4.694	**260.0**	1129.0	1469.7	2598.6	1134.9	1661.5	2796.4
5.000	263.9	1148.0	1449.0	2597.0	1154.5	1639.7	2794.2
5.506	**270.0**	1178.1	1415.9	2593.9	1185.2	1604.6	2789.9
6.000	275.6	1205.8	1384.6	2590.4	1213.7	1571.3	2785.0
6.420	**280.0**	1228.3	1358.7	2587.0	1236.8	1543.6	2780.4
7.000	285.8	1258.0	1323.9	2581.8	1267.4	1506.0	2773.5
7.446	**290.0**	1279.8	1297.7	2577.5	1290.0	1477.6	2767.6
8.000	295.0	1306.0	1265.7	2571.7	1317.1	1442.8	2759.9
8.593	**300.0**	1333.0	1232.0	2565.0	1345.1	1406.0	2751.0
9.000	303.3	1351.0	1209.2	2560.1	1363.7	1380.9	2744.6
9.870	**310.0**	1388.1	1161.0	2549.1	1402.4	1327.6	2730.0
10.000	311.0	1393.5	1153.8	2547.3	1408.0	1319.7	2727.7
11.000	318.1	1434.2	1099.0	2533.2	1450.6	1258.7	2709.3

Table A1 *(Cont'd)*

p_s [MPa]	t_s [°C]	s_f [kJ/kg K]	s_{fg} [kJ/kg K]	s_g [kJ/kg K]	v_f [m^3/kg]	v_{fg} [m^3/kg]	v_g [m^3/kg]
0.361	**140.0**	1.739	5.189	6.928	0.001 080	0.507 413	0.5085
0.400	143.6	1.776	5.118	6.894	0.001 084	0.461 141	0.4622
0.450	147.9	1.820	5.034	6.855	0.001 088	0.412 666	0.4138
0.476	**150.0**	1.842	4.994	6.836	0.001 091	0.391 357	0.3924
0.500	151.8	1.860	4.959	6.819	0.001 093	0.373 583	0.3747
0.600	158.8	1.931	4.827	6.758	0.001 101	0.314 373	0.3155
0.618	**160.0**	1.943	4.805	6.748	0.001 102	0.305 654	0.3068
0.700	165.0	1.992	4.713	6.705	0.001 108	0.271 573	0.2727
0.792	**170.0**	2.042	4.621	6.663	0.001 114	0.241 439	0.2426
0.800	170.4	2.046	4.614	6.660	0.001 115	0.239 142	0.2403
0.900	175.4	2.094	4.525	6.619	0.001 121	0.213 691	0.2148
1.000	179.9	2.138	4.445	6.583	0.001 127	0.193 165	0.1943
1.003	**180.0**	2.139	4.443	6.582	0.001 128	0.192 673	0.1938
1.255	**190.0**	2.236	4.268	6.504	0.001 142	0.155 174	0.1563
1.500	198.3	2.315	4.126	6.441	0.001 154	0.130 502	0.1317
1.555	**200.0**	2.331	4.097	6.428	0.001 156	0.126 004	0.1272
1.908	**210.0**	2.425	3.929	6.354	0.001 173	0.103 066	0.1042
2.000	212.4	2.447	3.890	6.337	0.001 177	0.098 360	0.099 54
2.320	**220.0**	2.518	3.764	6.282	0.001 190	0.084 848	0.086 04
2.500	223.9	2.554	3.699	6.254	0.001 197	0.078 708	0.079 91
2.798	**230.0**	2.610	3.601	6.211	0.001 209	0.070 241	0.071 45
3.000	233.8	2.646	3.538	6.184	0.001 216	0.065 410	0.066 63
3.348	**240.0**	2.702	3.439	6.141	0.001 229	0.058 425	0.059 65
3.500	242.5	2.725	3.398	6.123	0.001 235	0.055 791	0.057 03
3.978	**250.0**	2.794	3.277	6.071	0.001 251	0.048 786	0.050 04
4.000	250.3	2.797	3.272	6.069	0.001 252	0.048 497	0.049 75
4.500	257.4	2.861	3.158	6.019	0.001 269	0.042 768	0.044 04
4.694	**260.0**	2.885	3.116	6.001	0.001 276	0.040 858	0.042 13
5.000	263.9	2.921	3.053	5.974	0.001 286	0.038 143	0.039 43
5.506	**270.0**	2.976	2.954	5.930	0.001 303	0.034 285	0.035 59
6.000	275.6	3.027	2.864	5.891	0.001 319	0.031 119	0.032 44
6.420	**280.0**	3.068	2.790	5.859	0.001 332	0.028 794	0.030 13
7.000	285.8	3.122	2.694	5.816	0.001 351	0.026 022	0.027 37
7.446	**290.0**	3.161	2.624	5.785	0.001 366	0.024 169	0.025 54
8.000	295.0	3.208	2.540	5.747	0.001 384	0.022 141	0.023 53
8.593	**300.0**	3.255	2.453	5.708	0.001 404	0.020 245	0.021 65
9.000	303.3	3.287	2.395	5.682	0.001 418	0.019 078	0.020 50
9.870	**310.0**	3.351	2.277	5.628	0.001 448	0.016 886	0.018 33
10.000	311.0	3.361	2.259	5.620	0.001 453	0.016 589	0.018 04
11.000	318.1	3.430	2.129	5.560	0.001 489	0.014 517	0.016 01

Table A1 *(Cont'd)*

p_s [MPa]	t_s [°C]	u_f [kJ/kg]	u_{fg} [kJ/kg]	u_g [kJ/kg]	h_f [kJ/kg]	h_{fg} [kJ/kg]	h_g [kJ/kg]
11.289	**320.0**	1445.7	1083.2	2528.9	1462.6	1241.1	2703.7
12.000	324.7	1473.5	1044.3	2517.8	1491.8	1197.4	2689.2
12.863	**330.0**	1506.4	996.7	2503.1	1526.5	1143.6	2670.2
13.000	330.8	1511.6	989.0	2500.6	1532.0	1135.0	2667.0
14.000	336.6	1549.1	932.4	2481.4	1571.6	1070.7	2642.4
14.605	**340.0**	1571.5	897.2	2468.7	1595.5	1030.7	2626.2
15.000	342.1	1586.1	873.8	2459.9	1611.0	1004.0	2615.0
16.000	347.3	1623.2	812.8	2436.0	1650.5	934.3	2584.9
16.535	**350.0**	1643.2	779.0	2422.2	1671.9	895.7	2567.7
17.000	352.3	1661.6	747.7	2409.3	1691.7	859.9	2551.6
18.000	357.0	1701.7	677.2	2378.9	1734.8	779.1	2513.9
18.675	**360.0**	1728.8	627.1	2355.8	1764.2	721.3	2485.4
19.000	361.4	1742.1	601.7	2343.8	1778.7	692.0	2470.6
20.000	365.7	1785.7	515.1	2300.8	1826.5	591.9	2418.4
21.000	369.8	1840.0	402.1	2242.1	1886.3	461.3	2347.6
21.054	**370.0**	1843.6	394.5	2238.1	1890.2	452.6	2342.8
22.000	373.7	1952.4	161.3	2113.6	2011.1	184.5	2195.6
22.120	374.2	2037.3	0.0	2037.3	2107.4	0.0	2107.4

Table A1 *(Cont'd)*

p_s [MPa]	t_s [°C]	s_f [kJ/kg K]	s_{fg} [kJ/kg K]	s_g [kJ/kg K]	v_f [m^3/kg]	v_{fg} [m^3/kg]	v_g [m^3/kg]
11.289	**320.0**	3.450	2.092	5.542	0.001 500	0.013 980	0.015 48
12.000	324.7	3.497	2.003	5.500	0.001 527	0.012 756	0.014 28
12.863	**330.0**	3.553	1.896	5.449	0.001 561	0.011 428	0.012 99
13.000	330.8	3.562	1.879	5.441	0.001 567	0.011 230	0.012 80
14.000	336.6	3.624	1.756	5.380	0.001 611	0.009 884	0.011 50
14.605	**340.0**	3.662	1.681	5.343	0.001 639	0.009 142	0.010 78
15.000	342.1	3.686	1.632	5.318	0.001 658	0.008 682	0.010 34
16.000	347.3	3.747	1.506	5.253	0.001 710	0.007 597	0.009 308
16.535	**350.0**	3.780	1.438	5.218	0.001 741	0.007 058	0.008 799
17.000	352.3	3.811	1.375	5.186	0.001 770	0.006 601	0.008 371
18.000	357.0	3.877	1.236	5.113	0.001 840	0.005 658	0.007 498
18.675	**360.0**	3.921	1.139	5.060	0.001 896	0.005 044	0.006 940
19.000	361.4	3.943	1.090	5.033	0.001 926	0.004 751	0.006 678
20.000	365.7	4.015	0.926	4.941	0.002 037	0.003 840	0.005 877
21.000	369.8	4.105	0.718	4.822	0.002 202	0.002 822	0.005 023
21.054	**370.0**	4.111	0.704	4.814	0.002 214	0.002 759	0.004 973
22.000	373.7	4.295	0.285	4.580	0.002 671	0.001 056	0.003 728
22.120	374.2	4.443	0.000	4.443	0.003 170	0.000 000	0.003 170

Table A2 *Superheat table for water substance*

p/[MPa]				0.000 611				
t/[°C]	**0.0**	50	100	150	200	250	300	350
h/[kJ/kg]	2501.6	2594.7	2688.7	2783.8	2880.1	2977.7	3076.8	3177.5
s/[kJ/kg K]	9.157	9.470	9.741	9.980	10.195	10.391	10.572	10.740
v/[m^3/kg]	206.163	243.945	281.715	319.476	357.233	394.988	432.741	470.494
p/[MPa]				**0.001**				
t/[°C]	7.0	50	100	150	200	250	300	350
h/[kJ/kg]	2514.4	2594.6	2688.6	2783.7	2880.1	2977.7	3076.8	3177.5
s/[kJ/kg K]	8.977	9.243	9.514	9.753	9.968	10.164	10.345	10.513
v/[m^3/kg]	129.209	149.093	172.187	195.272	218.352	241.431	264.508	287.585
p/[MPa]				0.001 23				
t/[°C]	**10.0**	50	100	150	200	250	300	350
h/[kJ/kg]	2519.9	2594.5	2688.6	2783.7	2880.1	2977.7	3076.8	3177.5
s/[kJ/kg K]	8.902	9.148	9.419	9.658	9.873	10.070	10.251	10.419
v/[m^3/kg]	106.430	121.503	140.328	159.144	177.956	196.766	215.574	234.382
p/[MPa]				**0.002**				
t/[°C]	17.5	50	100	150	200	250	300	350
h/[kJ/kg]	2533.6	2594.4	2688.5	2783.7	2880.0	2977.7	3076.8	3177.5
s/[kJ/kg K]	8.725	8.923	9.193	9.433	9.648	9.844	10.025	10.193
v/[m^3/kg]	67.006	74.524	86.080	97.628	109.171	120.711	132.251	143.790
p/[MPa]				0.002 34				
t/[°C]	**20.0**	50	100	150	200	250	300	350
h/[kJ/kg]	2538.2	2594.3	2688.5	2783.6	2880.0	2977.7	3076.8	3177.5
s/[kJ/kg K]	8.668	8.851	9.122	9.361	9.576	9.772	9.953	10.122
v/[m^3/kg]	57.838	63.783	73.678	83.563	93.444	103.323	113.200	123.077
p/[MPa]				**0.003**				
t/[°C]	24.1	50	100	150	200	250	300	350
h/[kJ/kg]	2545.6	2594.2	2688.4	2783.6	2880.0	2977.7	3076.8	3177.4
s/[kJ/kg K]	8.578	8.735	9.006	9.245	9.461	9.657	9.838	10.006
v/[m^3/kg]	45.667	49.668	57.378	65.080	72.777	80.471	88.165	95.858
p/[MPa]				**0.004**				
t/[°C]	29.0	50	100	150	200	250	300	350
h/[kJ/kg]	2554.5	2593.9	2688.3	2783.5	2879.9	2977.6	3076.8	3177.4
s/[kJ/kg K]	8.475	8.602	8.873	9.113	9.328	9.524	9.705	9.873
v/[m^3/kg]	34.802	37.240	43.027	48.806	54.580	60.351	66.122	71.892
p/[MPa]				0.004 24				
t/[°C]	**30.0**	50	100	150	200	250	300	350
h/[kJ/kg]	2556.4	2593.9	2688.2	2783.5	2879.9	2977.6	3076.8	3177.4
s/[kJ/kg K]	8.455	8.574	8.846	9.085	9.301	9.497	9.678	9.846
v/[m^3/kg]	32.929	35.117	40.576	46.026	51.472	56.915	62.357	67.799
p/[MPa]				**0.005**				
t/[°C]	32.9	50	100	150	200	250	300	350
h/[kJ/kg]	2561.6	2593.7	2688.1	2783.4	2879.9	2977.6	3076.7	3177.4
s/[kJ/kg K]	8.396	8.498	8.770	9.009	9.225	9.421	9.602	9.770
v/[m^3/kg]	28.194	29.783	34.417	39.041	43.661	48.280	52.897	57.513
p/[MPa]				**0.006**				
t/[°C]	36.2	50	100	150	200	250	300	350
h/[kJ/kg]	2567.5	2593.5	2688.0	2783.4	2879.8	2977.6	3076.7	3177.4
s/[kJ/kg K]	8.331	8.413	8.685	8.925	9.141	9.337	9.518	9.686
v/[m^3/kg]	23.741	24.812	28.676	32.532	36.383	40.232	44.079	47.927

Table A2 *(Cont'd)*

p/[MPa]	**0.007**							
t/[°C]	39.0	50	100	150	200	250	300	350
h/[kJ/kg]	2572.6	2593.3	2687.9	2783.3	2879.8	2977.5	3076.7	3177.4
s/[kJ/kg K]	8.277	8.342	8.614	8.854	9.069	9.266	9.447	9.615
v/[m^3/kg]	20.531	21.261	24.576	27.882	31.184	34.483	37.781	41.079

p/[MPa]	0.007 38							
t/[°C]	**40.0**	50	100	150	200	250	300	350
h/[kJ/kg]	2574.4	2593.2	2687.8	2783.3	2879.8	2977.5	3076.7	3177.4
s/[kJ/kg K]	8.258	8.318	8.590	8.830	9.045	9.242	9.423	9.591
v/[m^3/kg]	19.546	20.177	23.325	26.463	29.597	32.729	35.860	38.990

p/[MPa]	**0.008**							
t/[°C]	41.5	50	100	150	200	250	300	350
h/[kJ/kg]	2577.1	2593.1	2687.8	2783.2	2879.7	2977.5	3076.7	3177.3
s/[kJ/kg K]	8.230	8.280	8.552	8.792	9.008	9.204	9.385	9.553
v/[m^3/kg]	18.105	18.598	21.501	24.395	27.284	30.172	33.058	35.944

p/[MPa]	**0.009**							
t/[°C]	43.8	50	100	150	200	250	300	350
h/[kJ/kg]	2581.1	2592.9	2687.6	2783.2	2879.7	2977.5	3076.6	3177.3
s/[kJ/kg K]	8.188	8.225	8.497	8.738	8.953	9.150	9.331	9.499
v/[m^3/kg]	16.204	16.526	19.109	21.682	24.251	26.818	29.384	31.949

p/[MPa]	**0.01**							
t/[°C]	45.8	50	100	150	200	250	300	350
h/[kJ/kg]	2584.8	2592.7	2687.5	2783.1	2879.6	2977.4	3076.6	3177.3
s/[kJ/kg K]	8.151	8.176	8.449	8.689	8.905	9.101	9.282	9.450
v/[m^3/kg]	14.675	14.869	17.195	19.512	21.825	24.136	26.445	28.754

p/[MPa]	0.0123							
t/[°C]	**50.0**	100	150	200	250	300	350	400
h/[kJ/kg]	2592.2	2687.2	2782.9	2879.5	2977.4	3076.6	3177.3	3279.5
s/[kJ/kg K]	8.078	8.351	8.592	8.807	9.004	9.185	9.354	9.511
v/[m^3/kg]	12.046	13.935	15.815	17.691	19.565	21.437	23.309	25.181

p/[MPa]	**0.015**							
t/[°C]	54.0	100	150	200	250	300	350	400
h/[kJ/kg]	2599.2	2686.9	2782.7	2879.4	2977.3	3076.5	3177.2	3279.5
s/[kJ/kg K]	8.009	8.260	8.501	8.717	8.914	9.095	9.263	9.421
v/[m^3/kg]	10.023	11.455	13.003	14.546	16.088	17.628	19.168	20.707

p/[MPa]	0.0199							
t/[°C]	**60.0**	100	150	200	250	300	350	400
h/[kJ/kg]	2609.7	2686.3	2782.3	2879.2	2977.1	3076.4	3177.1	3279.4
s/[kJ/kg K]	7.911	8.128	8.369	8.586	8.782	8.964	9.132	9.290
v/[m^3/kg]	7.679	8.619	9.787	10.951	12.112	13.272	14.432	15.591

p/[MPa]	**0.02**							
t/[°C]	60.1	100	150	200	250	300	350	400
h/[kJ/kg]	2609.9	2686.3	2782.3	2879.2	2977.1	3076.4	3177.1	3279.4
s/[kJ/kg K]	7.909	8.126	8.368	8.584	8.781	8.962	9.130	9.288
v/[m^3/kg]	7.650	8.585	9.748	10.907	12.064	13.219	14.374	15.529

p/[MPa]	**0.03**							
t/[°C]	69.1	100	150	200	250	300	350	400
h/[kJ/kg]	2625.4	2685.1	2781.6	2878.7	2976.8	3076.1	3176.9	3279.3
s/[kJ/kg K]	7.770	7.936	8.179	8.396	8.593	8.774	8.943	9.101
v/[m^3/kg]	5.229	5.714	6.493	7.268	8.040	8.811	9.581	10.351

Table A2 *(Cont'd)*

p/[MPa]				0.0312				
t/[°C]	**70.0**	100	150	200	250	300	350	400
h/[kJ/kg]	2626.9	2684.9	2781.5	2878.6	2976.8	3076.1	3176.9	3279.3
s/[kJ/kg K]	7.756	7.918	8.161	8.378	8.575	8.757	8.925	9.083
v/[m^3/kg]	5.046	5.500	6.250	6.996	7.740	8.482	9.224	9.965
p/[MPa]				**0.04**				
t/[°C]	75.9	100	150	200	250	300	350	400
h/[kJ/kg]	2636.9	2683.8	2780.9	2878.2	2976.5	3075.9	3176.8	3279.1
s/[kJ/kg K]	7.671	7.801	8.045	8.262	8.460	8.641	8.810	8.968
v/[m^3/kg]	3.993	4.279	4.866	5.448	6.028	6.606	7.185	7.762
p/[MPa]				0.0474				
t/[°C]	**80.0**	100	150	200	250	300	350	400
h/[kJ/kg]	2643.8	2682.9	2780.3	2877.9	2976.2	3075.7	3176.6	3279.0
s/[kJ/kg K]	7.613	7.721	7.966	8.184	8.382	8.563	8.732	8.890
v/[m^3/kg]	3.409	3.610	4.107	4.599	5.090	5.579	6.067	6.556
p/[MPa]				**0.05**				
t/[°C]	81.3	100	150	200	250	300	350	400
h/[kJ/kg]	2646.0	2682.6	2780.1	2877.7	2976.1	3075.7	3176.6	3279.0
s/[kJ/kg K]	7.595	7.695	7.941	8.159	8.356	8.538	8.707	8.865
v/[m^3/kg]	3.240	3.418	3.889	4.356	4.821	5.284	5.747	6.209
p/[MPa]				**0.06**				
t/[°C]	86.0	100	150	200	250	300	350	400
h/[kJ/kg]	2653.6	2681.3	2779.4	2877.3	2975.8	3075.4	3176.4	3278.8
s/[kJ/kg K]	7.533	7.608	7.855	8.074	8.272	8.454	8.622	8.781
v/[m^3/kg]	2.732	2.844	3.238	3.628	4.016	4.402	4.788	5.174
p/[MPa]				**0.07**				
t/[°C]	90.0	100	150	200	250	300	350	400
h/[kJ/kg]	2660.1	2680.0	2778.6	2876.8	2975.5	3075.2	3176.2	3278.7
s/[kJ/kg K]	7.480	7.535	7.783	8.002	8.200	8.382	8.551	8.709
v/[m^3/kg]	2.365	2.434	2.773	3.108	3.441	3.772	4.103	4.434
p/[MPa]				0.0701				
t/[°C]	**90.0**	100	150	200	250	300	350	400
h/[kJ/kg]	2660.1	2680.0	2778.6	2876.8	2975.5	3075.2	3176.2	3278.7
s/[kJ/kg K]	7.480	7.534	7.782	8.001	8.199	8.381	8.550	8.709
v/[m^3/kg]	2.361	2.430	2.769	3.103	3.436	3.767	4.097	4.427
p/[MPa]				**0.08**				
t/[°C]	93.5	100	150	200	250	300	350	400
h/[kJ/kg]	2665.8	2678.8	2777.8	2876.3	2975.2	3075.0	3176.0	3278.5
s/[kJ/kg K]	7.435	7.470	7.720	7.939	8.138	8.320	8.489	8.647
v/[m^3/kg]	2.087	2.126	2.425	2.718	3.010	3.300	3.590	3.879
p/[MPa]				**0.09**				
t/[°C]	96.7	100	150	200	250	300	350	400
h/[kJ/kg]	2670.9	2677.5	2777.1	2875.8	2974.8	3074.7	3175.8	3278.4
s/[kJ/kg K]	7.395	7.413	7.664	7.884	8.083	8.266	8.435	8.593
v/[m^3/kg]	1.869	1.887	2.153	2.415	2.674	2.933	3.190	3.448
p/[MPa]				**0.1**				
t/[°C]	99.6	100	150	200	250	300	350	400
h/[kJ/kg]	2675.4	2676.2	2776.3	2875.4	2974.5	3074.5	3175.6	3278.2
s/[kJ/kg K]	7.360	7.362	7.614	7.835	8.034	8.217	8.386	8.544
v/[m^3/kg]	1.694	1.696	1.936	2.172	2.406	2.639	2.871	3.102

Table A2 *(Cont'd)*

p/[MPa]				0.101 325				
t/[°C]	**100.0**	150	200	250	300	350	400	450
h/[kJ/kg]	2676.0	2776.2	2875.3	2974.5	3074.4	3175.6	3278.2	3382.3
s/[kJ/kg K]	7.355	7.607	7.829	8.028	8.211	8.380	8.538	8.687
v/[m^3/kg]	1.673	1.911	2.144	2.375	2.604	2.833	3.062	3.290
p/[MPa]				0.143				
t/[°C]	**110.0**	150	200	250	300	350	400	450
h/[kJ/kg]	2691.3	2773.0	2873.3	2973.1	3073.4	3174.8	3277.6	3381.8
s/[kJ/kg K]	7.239	7.442	7.666	7.866	8.049	8.219	8.378	8.527
v/[m^3/kg]	1.210	1.346	1.513	1.677	1.840	2.002	2.164	2.326
p/[MPa]				**0.15**				
t/[°C]	111.4	150	200	250	300	350	400	450
h/[kJ/kg]	2693.4	2772.5	2872.9	2972.9	3073.3	3174.7	3277.5	3381.7
s/[kJ/kg K]	7.223	7.419	7.644	7.845	8.028	8.198	8.356	8.506
v/[m^3/kg]	1.159	1.285	1.444	1.601	1.757	1.912	2.067	2.222
p/[MPa]				0.199				
t/[°C]	**120.0**	150	200	250	300	350	400	450
h/[kJ/kg]	2706.0	2768.6	2870.6	2971.3	3072.1	3173.8	3276.7	3381.1
s/[kJ/kg K]	7.129	7.283	7.511	7.713	7.897	8.067	8.226	8.376
v/[m^3/kg]	0.8915	0.9667	1.088	1.208	1.326	1.443	1.561	1.678
p/[MPa]				**0.2**				
t/[°C]	120.2	150	200	250	300	350	400	450
h/[kJ/kg]	2706.3	2768.5	2870.5	2971.2	3072.1	3173.8	3276.7	3381.1
s/[kJ/kg K]	7.127	7.279	7.507	7.710	7.894	8.064	8.223	8.372
v/[m^3/kg]	0.8854	0.9595	1.080	1.199	1.316	1.433	1.549	1.665
p/[MPa]				**0.25**				
t/[°C]	127.4	150	200	250	300	350	400	450
h/[kJ/kg]	2716.4	2764.5	2868.0	2969.6	3070.9	3172.8	3275.9	3380.4
s/[kJ/kg K]	7.052	7.169	7.400	7.604	7.789	7.960	8.119	8.269
v/[m^3/kg]	0.7184	0.7641	0.8620	0.9574	1.052	1.145	1.238	1.332
p/[MPa]				0.270				
t/[°C]	**130.0**	150	200	250	300	350	400	450
h/[kJ/kg]	2719.9	2762.9	2867.0	2968.9	3070.4	3172.4	3275.6	3380.2
s/[kJ/kg K]	7.026	7.130	7.363	7.567	7.753	7.923	8.083	8.233
v/[m^3/kg]	0.6681	0.7058	0.7969	0.8854	0.9728	1.060	1.146	1.232
p/[MPa]				**0.3**				
t/[°C]	133.5	150	200	250	300	350	400	450
h/[kJ/kg]	2724.7	2760.4	2865.5	2967.9	3069.7	3171.9	3275.2	3379.8
s/[kJ/kg K]	6.991	7.077	7.312	7.518	7.703	7.874	8.034	8.184
v/[m^3/kg]	0.6056	0.6337	0.7164	0.7964	0.8753	0.9535	1.031	1.109
p/[MPa]				**0.35**				
t/[°C]	138.9	150	200	250	300	350	400	450
h/[kJ/kg]	2731.6	2756.3	2863.0	2966.2	3068.4	3170.9	3274.4	3379.2
s/[kJ/kg K]	6.939	6.998	7.237	7.444	7.631	7.802	7.962	8.112
v/[m^3/kg]	0.5240	0.5406	0.6123	0.6814	0.7493	0.8166	0.8835	0.9501
p/[MPa]				0.361				
t/[°C]	**140.0**	150	200	250	300	350	400	450
h/[kJ/kg]	2733.1	2755.3	2862.4	2965.8	3068.2	3170.7	3274.2	3379.0
s/[kJ/kg K]	6.928	6.982	7.221	7.429	7.616	7.787	7.947	8.097
v/[m^3/kg]	0.508	0.523	0.593	0.660	0.726	0.791	0.856	0.920

Table A2 *(Cont'd)*

p/[MPa]	**0.4**							
t/[°C]	143.6	150	200	250	300	350	400	450
h/[kJ/kg]	2737.6	2752.0	2860.4	2964.5	3067.2	3170.0	3273.6	3378.5
s/[kJ/kg K]	6.894	6.929	7.171	7.380	7.568	7.739	7.899	8.050
v/[m^3/kg]	0.4622	0.4707	0.5343	0.5952	0.6549	0.7139	0.7725	0.8309

p/[MPa]	**0.45**							
t/[°C]	147.9	150	200	250	300	350	400	450
h/[kJ/kg]	2742.9	2747.7	2857.8	2962.8	3066.0	3169.1	3272.9	3377.9
s/[kJ/kg K]	6.855	6.866	7.112	7.323	7.512	7.684	7.844	7.995
v/[m^3/kg]	0.4138	0.4162	0.4735	0.5281	0.5814	0.6340	0.6862	0.7382

p/[MPa]	0.476							
t/[°C]	**150.0**	200	250	300	350	400	450	500
h/[kJ/kg]	2745.4	2856.4	2961.9	3065.3	3168.6	3272.5	3377.5	3484.0
s/[kJ/kg K]	6.836	7.084	7.296	7.485	7.658	7.818	7.968	8.111
v/[m^3/kg]	0.3924	0.4470	0.4988	0.5493	0.5991	0.6485	0.6977	0.7468

p/[MPa]	**0.5**							
t/[°C]	151.8	200	250	300	350	400	450	500
h/[kJ/kg]	2747.5	2855.1	2961.1	3064.8	3168.1	3272.1	3377.2	3483.8
s/[kJ/kg K]	6.819	7.059	7.272	7.461	7.634	7.795	7.945	8.088
v/[m^3/kg]	0.3747	0.4250	0.4744	0.5226	0.5701	0.6172	0.6640	0.7108

p/[MPa]	**0.6**							
t/[°C]	158.8	200	250	300	350	400	450	500
h/[kJ/kg]	2755.5	2849.7	2957.6	3062.3	3166.2	3270.6	3376.0	3482.7
s/[kJ/kg K]	6.758	6.966	7.183	7.374	7.548	7.709	7.860	8.003
v/[m^3/kg]	0.3155	0.3520	0.3939	0.4344	0.4742	0.5136	0.5528	0.5918

p/[MPa]	0.618							
t/[°C]	**160.0**	200	250	300	350	400	450	500
h/[kJ/kg]	2756.7	2848.8	2956.9	3061.8	3165.9	3270.3	3375.7	3482.5
s/[kJ/kg K]	6.747	6.951	7.168	7.360	7.534	7.695	7.846	7.989
v/[m^3/kg]	0.3068	0.3414	0.3821	0.4215	0.4602	0.4985	0.5365	0.5745

p/[MPa]	**0.7**							
t/[°C]	165.0	200	250	300	350	400	450	500
h/[kJ/kg]	2762.0	2844.2	2954.0	3059.8	3164.3	3269.0	3374.7	3481.6
s/[kJ/kg K]	6.705	6.886	7.107	7.300	7.475	7.636	7.788	7.930
v/[m^3/kg]	0.2727	0.2999	0.3364	0.3714	0.4057	0.4396	0.4733	0.5069

p/[MPa]	0.792							
t/[°C]	**170.0**	200	250	300	350	400	450	500
h/[kJ/kg]	2767.1	2839.0	2950.7	3057.5	3162.5	3267.6	3373.5	3480.6
s/[kJ/kg K]	6.663	6.820	7.045	7.240	7.415	7.578	7.729	7.873
v/[m^3/kg]	0.2426	0.2636	0.2963	0.3275	0.3580	0.3881	0.4179	0.4477

p/[MPa]	**0.8**							
t/[°C]	170.4	200	250	300	350	400	450	500
h/[kJ/kg]	2767.5	2838.6	2950.4	3057.3	3162.4	3267.5	3373.4	3480.5
s/[kJ/kg K]	6.660	6.815	7.040	7.235	7.411	7.573	7.725	7.868
v/[m^3/kg]	0.2403	0.2608	0.2932	0.3241	0.3543	0.3842	0.4137	0.4432

p/[MPa]	**0.9**							
t/[°C]	175.4	200	250	300	350	400	450	500
h/[kJ/kg]	2772.1	2832.7	2946.8	3054.7	3160.5	3266.0	3372.1	3479.4
s/[kJ/kg K]	6.619	6.751	6.980	7.177	7.354	7.517	7.669	7.812
v/[m^3/kg]	0.2148	0.2303	0.2596	0.2874	0.3144	0.3410	0.3674	0.3936

Table A2 *(Cont'd)*

p/[MPa]				**1**				
t/[°C]	179.9	200	250	300	350	400	450	500
h/[kJ/kg]	2776.2	2826.8	2943.0	3052.1	3158.5	3264.4	3370.8	3478.3
s/[kJ/kg K]	6.583	6.692	6.926	7.125	7.303	7.467	7.619	7.763
v/[m^3/kg]	0.1943	0.2059	0.2327	0.2580	0.2824	0.3065	0.3303	0.3540

p/[MPa]				1.003				
t/[°C]	**180.0**	200	250	300	350	400	450	500
h/[kJ/kg]	2776.3	2826.6	2942.9	3052.1	3158.5	3264.4	3370.8	3478.3
s/[kJ/kg K]	6.582	6.691	6.925	7.124	7.302	7.465	7.618	7.761
v/[m^3/kg]	0.1938	0.2053	0.2321	0.2573	0.2817	0.3057	0.3294	0.3530

p/[MPa]				1.255				
t/[°C]	**190.0**	200	250	300	350	400	450	500
h/[kJ/kg]	2784.3	2810.9	2933.3	3045.5	3153.6	3260.4	3367.5	3475.5
s/[kJ/kg K]	6.504	6.561	6.807	7.012	7.192	7.357	7.511	7.655
v/[m^3/kg]	0.1563	0.1612	0.1835	0.2042	0.2240	0.2433	0.2625	0.2814

p/[MPa]				**1.5**				
t/[°C]	198.3	200	250	300	350	400	450	500
h/[kJ/kg]	2789.9	2794.7	2923.5	3038.9	3148.7	3256.6	3364.3	3472.8
s/[kJ/kg K]	6.441	6.451	6.710	6.921	7.104	7.271	7.425	7.570
v/[m^3/kg]	0.1317	0.1324	0.1520	0.1697	0.1865	0.2029	0.2191	0.2350

p/[MPa]				1.555				
t/[°C]	**200.0**	250	300	350	400	450	500	550
h/[kJ/kg]	2790.9	2921.3	3037.4	3147.6	3255.7	3363.6	3472.2	3581.9
s/[kJ/kg K]	6.428	6.690	6.902	7.087	7.253	7.408	7.553	7.691
v/[m^3/kg]	0.1272	0.1463	0.1635	0.1798	0.1956	0.2112	0.2266	0.2419

p/[MPa]				1.908				
t/[°C]	**210.0**	250	300	350	400	450	500	550
h/[kJ/kg]	2796.2	2906.4	3027.7	3140.5	3250.2	3359.0	3468.3	3578.5
s/[kJ/kg K]	6.354	6.573	6.795	6.984	7.153	7.309	7.455	7.593
v/[m^3/kg]	0.1042	0.1173	0.1319	0.1455	0.1586	0.1715	0.1842	0.1968

p/[MPa]				**2**				
t/[°C]	212.4	250	300	350	400	450	500	550
h/[kJ/kg]	2797.2	2902.4	3025.0	3138.6	3248.7	3357.8	3467.3	3577.6
s/[kJ/kg K]	6.337	6.545	6.770	6.960	7.130	7.286	7.432	7.571
v/[m^3/kg]	0.099 54	0.1114	0.1255	0.1386	0.1511	0.1634	0.1756	0.1876

p/[MPa]				2.320				
t/[°C]	**220.0**	250	300	350	400	450	500	550
h/[kJ/kg]	2799.9	2887.9	3015.8	3132.0	3243.6	3353.6	3463.7	3574.6
s/[kJ/kg K]	6.282	6.455	6.689	6.883	7.055	7.213	7.360	7.499
v/[m^3/kg]	0.086 04	0.094 60	0.1072	0.1187	0.1297	0.1404	0.1510	0.1614

p/[MPa]				**2.5**				
t/[°C]	223.9	250	300	350	400	450	500	550
h/[kJ/kg]	2800.9	2879.5	3010.4	3128.2	3240.7	3351.3	3461.7	3572.9
s/[kJ/kg K]	6.254	6.408	6.647	6.844	7.018	7.176	7.324	7.463
v/[m^3/kg]	0.079 91	0.086 99	0.098 93	0.1098	0.1200	0.1300	0.1399	0.1496

p/[MPa]				2.798				
t/[°C]	**230.0**	250	300	350	400	450	500	550
h/[kJ/kg]	2802.0	2865.0	3001.4	3121.9	3235.8	3347.3	3458.4	3570.1
s/[kJ/kg K]	6.211	6.334	6.583	6.785	6.961	7.120	7.269	7.409
v/[m^3/kg]	0.071 45	0.076 51	0.087 59	0.097 48	0.1068	0.1158	0.1247	0.1334

Table A2 *(Cont'd)*

p/[MPa]	**3**							
t/[°C]	233.8	250	300	350	400	450	500	550
h/[kJ/kg]	2802.3	2854.8	2995.1	3117.5	3232.5	3344.6	3456.2	3568.1
s/[kJ/kg K]	6.184	6.286	6.542	6.747	6.925	7.085	7.234	7.375
v/[m^3/kg]	0.066 63	0.070 55	0.081 16	0.090 53	0.099 31	0.1078	0.1161	0.1243

p/[MPa]	3.348							
t/[°C]	**240.0**	250	300	350	400	450	500	550
h/[kJ/kg]	2802.2	2836.5	2983.9	3109.9	3226.7	3340.0	3452.3	3564.8
s/[kJ/kg K]	6.141	6.207	6.476	6.687	6.868	7.030	7.180	7.321
v/[m^3/kg]	0.059 65	0.061 94	0.071 90	0.080 53	0.088 54	0.096 23	0.1037	0.1111

p/[MPa]	**3.5**							
t/[°C]	242.5	250	300	350	400	450	500	550
h/[kJ/kg]	2802.0	2828.1	2979.0	3106.5	3224.2	3338.0	3450.6	3563.4
s/[kJ/kg K]	6.123	6.173	6.449	6.663	6.844	7.007	7.158	7.299
v/[m^3/kg]	0.057 03	0.058 69	0.068 42	0.076 78	0.084 49	0.091 89	0.099 09	0.1062

p/[MPa]	3.978							
t/[°C]	**250.0**	300	350	400	450	500	550	600
h/[kJ/kg]	2900.4	2962.8	3095.6	3216.1	3331.5	3445.2	3558.8	3672.9
s/[kJ/kg K]	6.071	6.368	6.590	6.776	6.942	7.094	7.236	7.371
v/[m^3/kg]	0.050 04	0.059 21	0.066 85	0.073 81	0.080 43	0.086 84	0.093 13	0.099 33

p/[MPa]	**4**							
t/[°C]	250.3	300	350	400	450	500	550	600
h/[kJ/kg]	2800.3	2962.0	3095.1	3215.7	3331.2	3445.0	3558.6	3672.8
s/[kJ/kg K]	6.069	6.364	6.587	6.773	6.939	7.091	7.233	7.368
v/[m^3/kg]	0.049 75	0.058 83	0.066 45	0.073 38	0.079 96	0.086 34	0.092 60	0.098 76

p/[MPa]	**4.5**							
t/[°C]	257.4	300	350	400	450	500	550	600
h/[kJ/kg]	2797.7	2944.2	3083.3	3207.1	3324.4	3439.3	3553.8	3668.6
s/[kJ/kg K]	6.019	6.285	6.518	6.709	6.877	7.031	7.175	7.310
v/[m^3/kg]	0.044 04	0.051 34	0.058 40	0.064 72	0.070 68	0.076 43	0.082 04	0.087 57

p/[MPa]	4.694							
t/[°C]	**260.0**	300	350	400	450	500	550	600
h/[kJ/kg]	2796.4	2937.1	3078.6	3203.7	3321.7	3437.1	3551.9	3667.0
s/[kJ/kg K]	6.001	6.256	6.493	6.686	6.855	7.009	7.153	7.289
v/[m^3/kg]	0.042 13	0.048 84	0.055 73	0.061 85	0.067 60	0.073 14	0.078 55	0.083 86

p/[MPa]	**5**							
t/[°C]	263.9	300	350	400	450	500	550	600
h/[kJ/kg]	2794.2	2925.5	3071.2	3198.3	3317.5	3433.7	3549.0	3664.5
s/[kJ/kg K]	5.973	6.210	6.455	6.651	6.822	6.977	7.122	7.258
v/[m^3/kg]	0.039 43	0.045 30	0.051 94	0.057 79	0.063 25	0.068 49	0.073 60	0.078 62

p/[MPa]	5.506							
t/[°C]	**270.0**	300	350	400	450	500	550	600
h/[kJ/kg]	2789.9	2905.6	3058.5	3189.2	3310.4	3427.9	3544.1	3660.3
s/[kJ/kg K]	5.930	6.138	6.394	6.596	6.770	6.927	7.073	7.210
v/[m^3/kg]	0.035 59	0.040 27	0.046 59	0.052 06	0.057 11	0.061 93	0.066 62	0.071 21

p/[MPa]	**6**							
t/[°C]	275.6	300	350	400	450	500	550	600
h/[kJ/kg]	2785.0	2885.0	3045.8	3180.1	3303.5	3422.2	3539.3	3656.2
s/[kJ/kg K]	5.891	6.069	6.339	6.546	6.723	6.882	7.029	7.166
v/[m^3/kg]	0.032 44	0.036 15	0.042 22	0.047 38	0.052 10	0.056 59	0.060 94	0.065 18

Table A2 *(Cont'd)*

p/[MPa]				6.420				
t/[°C]	**280.0**	300	350	400	450	500	550	600
h/[kJ/kg]	2780.4	2866.5	3034.6	3172.3	3297.5	3417.3	3535.2	3652.7
s/[kJ/kg K]	5.859	6.012	6.293	6.506	6.685	6.846	6.994	7.132
v/[m^3/kg]	0.030 13	0.033 10	0.039 03	0.043 97	0.048 45	0.052 70	0.056 79	0.060 79
p/[MPa]				**7**				
t/[°C]	285.8	300	350	400	450	500	550	600
h/[kJ/kg]	2773.5	2839.4	3018.7	3161.2	3289.1	3410.6	3529.6	3647.9
s/[kJ/kg K]	5.816	5.933	6.233	6.454	6.637	6.799	6.948	7.088
v/[m^3/kg]	0.027 37	0.029 46	0.035 23	0.039 92	0.044 13	0.048 09	0.051 89	0.055 59
p/[MPa]				7.446				
t/[°C]	**290.0**	300	350	400	450	500	550	600
h/[kJ/kg]	2767.6	2816.9	3006.1	3152.6	3282.5	3405.3	3525.2	3644.2
s/[kJ/kg K]	5.785	5.872	6.189	6.415	6.601	6.766	6.916	7.056
v/[m^3/kg]	0.025 54	0.027 00	0.032 71	0.037 24	0.041 26	0.045 03	0.048 64	0.052 14
p/[MPa]				**8**				
t/[°C]	295.0	300	350	400	450	500	550	600
h/[kJ/kg]	2759.9	2786.8	2989.9	3141.6	3274.3	3398.8	3519.7	3639.5
s/[kJ/kg K]	5.747	5.794	6.135	6.369	6.560	6.726	6.878	7.019
v/[m^3/kg]	0.023 53	0.024 26	0.029 95	0.034 31	0.038 15	0.041 70	0.045 10	0.048 39
p/[MPa]				8.593				
t/[°C]	**300.0**	350	400	450	500	550	600	650
h/[kJ/kg]	2751.0	2971.9	3129.6	3265.4	3391.7	3513.9	3634.6	3754.9
s/[kJ/kg K]	5.708	6.079	6.323	6.517	6.686	6.839	6.982	7.116
v/[m^3/kg]	0.021 65	0.027 37	0.031 59	0.035 25	0.038 62	0.041 82	0.044 92	0.047 94
p/[MPa]				**9**				
t/[°C]	303.3	350	400	450	500	550	600	650
h/[kJ/kg]	2744.6	2959.0	3121.2	3259.2	3386.8	3509.8	3631.1	3752.0
s/[kJ/kg K]	5.682	6.041	6.292	6.489	6.660	6.814	6.957	7.092
v/[m^3/kg]	0.020 50	0.025 79	0.029 93	0.033 48	0.036 74	0.039 82	0.042 80	0.045 70
p/[MPa]				9.870				
t/[°C]	**310.0**	350	400	450	500	550	600	650
h/[kJ/kg]	2730.0	2930.2	3102.7	3245.7	3376.2	3501.1	3623.8	3745.7
s/[kJ/kg K]	5.628	5.961	6.227	6.432	6.607	6.764	6.908	7.044
v/[m^3/kg]	0.018 33	0.022 82	0.026 83	0.030 19	0.033 23	0.036 10	0.038 85	0.041 52
p/[MPa]				**10**				
t/[°C]	311.0	350	400	450	500	550	600	650
h/[kJ/kg]	2727.7	2925.8	3099.9	3243.6	3374.6	3499.8	3622.7	3744.7
s/[kJ/kg K]	5.620	5.949	6.218	6.424	6.599	6.756	6.901	7.037
v/[m^3/kg]	0.018 04	0.022 42	0.026 41	0.029 74	0.032 76	0.035 60	0.038 32	0.040 96
p/[MPa]				**11**				
t/[°C]	318.0	350	400	450	500	550	600	650
h/[kJ/kg]	2709.3	2889.6	3077.8	3227.7	3362.2	3489.7	3614.2	3737.5
s/[kJ/kg K]	5.560	5.857	6.148	6.363	6.543	6.703	6.850	6.987
v/[m^3/kg]	0.016 01	0.019 61	0.023 51	0.026 68	0.029 50	0.032 14	0.034 66	0.037 09
p/[MPa]				11.289				
t/[°C]	**320.0**	350	400	450	500	550	600	650
h/[kJ/kg]	2703.7	2878.4	3071.2	3223.0	3358.6	3486.8	3611.7	3735.4
s/[kJ/kg K]	5.542	5.830	6.129	6.346	6.528	6.688	6.836	6.973
v/[m^3/kg]	0.015 48	0.018 88	0.022 77	0.025 89	0.028 67	0.031 25	0.033 72	0.036 10

Table A2 *(Cont'd)*

p/[MPa]				12				
t/[°C]	324.6	350	400	450	500	550	600	650
h/[kJ/kg]	2689.2	2849.7	3054.8	3211.4	3349.6	3479.6	3605.7	3730.2
s/[kJ/kg K]	5.500	5.764	6.081	6.306	6.491	6.653	6.802	6.941
v/[m^3/kg]	0.014 28	0.017 21	0.021 08	0.024 12	0.026 79	0.029 26	0.031 60	0.033 87
p/[MPa]				12.863				
t/[°C]	**330.0**	350	400	450	500	550	600	650
h/[kJ/kg]	2670.2	2811.5	3034.1	3196.9	3338.6	3470.7	3598.3	3723.9
s/[kJ/kg K]	5.449	5.680	6.024	6.258	6.448	6.613	6.764	6.904
v/[m^3/kg]	0.012 99	0.015 38	0.019 28	0.022 22	0.024 78	0.027 13	0.029 35	0.031 49
p/[MPa]				13				
t/[°C]	330.8	350	400	450	500	550	600	650
h/[kJ/kg]	2667.0	2805.0	3030.7	3194.6	3336.8	3469.3	3597.1	3722.9
s/[kJ/kg K]	5.441	5.666	6.016	6.251	6.441	6.607	6.758	6.898
v/[m^3/kg]	0.012 80	0.015 10	0.019 02	0.021 94	0.024 49	0.026 82	0.029 02	0.031 14
p/[MPa]				14				
t/[°C]	336.6	350	400	450	500	550	600	650
h/[kJ/kg]	2642.4	2754.2	3005.6	3177.4	3323.8	3458.8	3588.5	3715.6
s/[kJ/kg K]	5.380	5.562	5.951	6.198	6.394	6.563	6.716	6.858
v/[m^3/kg]	0.011 50	0.013 21	0.017 23	0.020 08	0.022 51	0.024 72	0.026 80	0.028 80
p/[MPa]				14.605				
t/[°C]	**340.0**	350	400	450	500	550	600	650
h/[kJ/kg]	2626.2	2719.5	2989.7	3166.8	3315.8	3452.5	3583.2	3711.2
s/[kJ/kg K]	5.343	5.494	5.913	6.167	6.366	6.538	6.692	6.834
v/[m^3/kg]	0.010 78	0.012 14	0.016 26	0.019 07	0.021 44	0.023 60	0.025 61	0.027 54
p/[MPa]				15				
t/[°C]	342.1	350	400	450	500	550	600	650
h/[kJ/kg]	2615.0	2694.8	2979.1	3159.7	3310.6	3448.3	3579.8	3708.3
s/[kJ/kg K]	5.318	5.447	5.888	6.147	6.349	6.521	6.676	6.819
v/[m^3/kg]	0.010 34	0.011 46	0.015 66	0.018 45	0.020 80	0.022 91	0.024 88	0.026 77
p/[MPa]				16				
t/[°C]	347.3	350	400	450	500	550	600	650
h/[kJ/kg]	2584.9	2620.8	2951.3	3141.6	3297.1	3437.7	3571.0	3700.9
s/[kJ/kg K]	5.253	5.311	5.824	6.097	6.305	6.482	6.639	6.783
v/[m^3/kg]	0.009 308	0.009 764	0.014 28	0.017 03	0.019 29	0.021 32	0.023 20	0.024 99
p/[MPa]				16.535				
t/[°C]	**350.0**	400	450	500	550	600	650	700
h/[kJ/kg]	2567.7	2935.7	3131.7	3289.9	3432.0	3566.3	3696.9	3825.8
s/[kJ/kg K]	5.218	5.790	6.071	6.283	6.461	6.620	6.765	6.901
v/[m^3/kg]	0.008 799	0.013 60	0.016 33	0.018 56	0.020 55	0.022 39	0.024 13	0.025 81
p/[MPa]				17				
t/[°C]	352.3	400	450	500	550	600	650	700
h/[kJ/kg]	2551.6	2921.7	3123.1	3283.5	3427.0	3562.2	3693.5	3822.8
s/[kJ/kg K]	5.185	5.760	6.049	6.264	6.444	6.603	6.749	6.886
v/[m^3/kg]	0.008 371	0.013 03	0.015 76	0.017 97	0.019 92	0.021 72	0.023 43	0.025 07
p/[MPa]				18				
t/[°C]	357.0	400	450	500	550	600	650	700
h/[kJ/kg]	2513.9	2890.3	3104.0	3269.6	3416.1	3553.4	3686.1	3816.5
s/[kJ/kg K]	5.113	5.695	6.002	6.223	6.407	6.569	6.717	6.854
v/[m^3/kg]	0.007 498	0.011 91	0.014 64	0.016 79	0.018 67	0.020 40	0.022 04	0.023 60

Table A2 *(Cont'd)*

p/[MPa]				18.675				
t/[°C]	**360.0**	400	450	500	550	600	650	700
h/[kJ/kg]	2485.4	2867.9	3090.8	3260.1	3408.7	3547.4	3681.0	3812.2
s/[kJ/kg K]	5.060	5.650	5.970	6.197	6.383	6.546	6.695	6.834
v/[m^3/kg]	0.006 940	0.011 21	0.013 94	0.016 06	0.017 90	0.019 59	0.021 18	0.022 70
p/[MPa]				**19**				
t/[°C]	361.4	400	450	500	550	600	650	700
h/[kJ/kg]	2470.6	2856.7	3084.4	3255.4	3405.2	3544.5	3678.6	3810.2
s/[kJ/kg K]	5.033	5.628	5.955	6.184	6.372	6.536	6.685	6.824
v/[m^3/kg]	0.006 678	0.010 89	0.013 62	0.015 73	0.017 55	0.019 22	0.020 79	0.022 29
p/[MPa]				**20**				
t/[°C]	365.7	400	450	500	550	600	650	700
h/[kJ/kg]	2418.4	2820.5	3064.3	3241.1	3394.1	3535.5	3671.1	3803.8
s/[kJ/kg K]	4.941	5.559	5.909	6.146	6.337	6.504	6.655	6.795
v/[m^3/kg]	0.005 877	0.009 947	0.012 71	0.014 77	0.016 55	0.018 16	0.019 67	0.021 11
p/[MPa]				**21**				
t/[°C]	369.8	400	450	500	550	600	650	700
h/[kJ/kg]	2347.6	2781.3	3043.6	3226.5	3382.9	3526.5	3663.6	3797.5
s/[kJ/kg K]	4.822	5.486	5.863	6.108	6.304	6.474	6.626	6.768
v/[m^3/kg]	0.005 023	0.009 071	0.011 87	0.013 91	0.015 64	0.017 20	0.018 66	0.020 04
p/[MPa]				21.054				
t/[°C]	**370.0**	400	450	500	550	600	650	700
h/[kJ/kg]	2342.8	2779.1	3042.4	3225.7	3382.3	3526.0	3663.2	3797.1
s/[kJ/kg K]	4.814	5.482	5.861	6.106	6.303	6.472	6.625	6.766
v/[m^3/kg]	0.004 973	0.009 026	0.011 83	0.013 86	0.015 59	0.017 15	0.018 61	0.019 99
p/[MPa]				**22**				
t/[°C]	373.7	400	450	500	550	600	650	700
h/[kJ/kg]	2195.6	2738.8	3022.3	3211.7	3371.6	3517.4	3656.1	3791.1
s/[kJ/kg K]	4.580	5.410	5.818	6.072	6.272	6.444	6.599	6.741
v/[m^3/kg]	0.003 728	0.008 251	0.011 11	0.013 12	0.014 81	0.016 33	0.017 74	0.019 07
p/[MPa]				22.120				
t/[°C]	374.2	400	450	500	550	600	650	700
h/[kJ/kg]	2107.4	2733.4	3019.7	3209.9	3370.2	3516.3	3655.2	3790.3
s/[kJ/kg K]	4.443	5.401	5.813	6.067	6.268	6.441	6.595	6.738
v/[m^3/kg]	0.003 170	0.008 156	0.011 03	0.013 03	0.014 72	0.016 23	0.017 63	0.018 96

Table A3 *Properties of water substance at supercritical pressures*

p/[MPa]				22.12			
t/[°C]	0.0	50	100	150	200	250	300
h/[kJ/kg]	22.2	228.2	435.7	645.8	861.4	1087.0	1332.8
s/[kJ/kg K]	0.001	0.693	1.290	1.818	2.300	2.753	3.201
v/[m^3/kg]	0.000 989	0.001 003	0.001 033	0.001 077	0.001 137	0.001 222	0.001 354
p/[MPa]				**30**			
t/[°C]	0.0	50	100	150	200	250	300
h/[kJ/kg]	30.0	235.0	441.6	650.9	865.2	1088.4	1328.7
s/[kJ/kg K]	0.001	0.690	1.284	1.811	2.289	2.737	3.176
v/[m^3/kg]	0.000 986	0.000 999	0.001 029	0.001 072	0.001 130	0.001 211	0.001 332
p/[MPa]				**40**			
t/[°C]	0.0	50	100	150	200	250	300
h/[kJ/kg]	39.7	243.5	449.2	657.4	870.2	1090.8	1325.4
s/[kJ/kg K]	0.000	0.685	1.277	1.801	2.276	2.719	3.147
v/[m^3/kg]	0.000 981	0.000 995	0.001 024	0.001 066	0.001 122	0.001 198	0.001 308
p/[MPa]				**50**			
t/[°C]	0.0	50	100	150	200	250	300
h/[kJ/kg]	49.3	251.9	456.8	664.1	875.4	1093.6	1323.7
s/[kJ/kg K]	0.000	0.681	1.270	1.791	2.263	2.701	3.121
v/[m^3/kg]	0.000 977	0.000 991	0.001 020	0.001 060	0.001 114	0.001 187	0.001 287
p/[MPa]				**60**			
t/[°C]	0.0	50	100	150	200	250	300
h/[kJ/kg]	58.8	260.4	464.5	670.7	880.8	1096.9	1323.2
s/[kJ/kg K]	−0.001	0.676	1.263	1.782	2.251	2.685	3.098
v/[m^3/kg]	0.000 972	0.000 988	0.001 016	0.001 055	0.001 107	0.001 176	0.001 270
p/[MPa]				**70**			
t/[°C]	0.0	50	100	150	200	250	300
h/[kJ/kg]	68.2	268.8	472.1	677.5	886.3	1100.5	1323.6
s/[kJ/kg K]	−0.002	0.672	1.257	1.773	2.239	2.670	3.077
v/[m^3/kg]	0.000 968	0.000 984	0.001 012	0.001 050	0.001 101	0.001 166	0.001 254
p/[MPa]				**80**			
t/[°C]	0.0	50	100	150	200	250	300
h/[kJ/kg]	77.5	277.2	479.7	684.3	891.9	1104.4	1324.7
s/[kJ/kg K]	−0.004	0.667	1.250	1.764	2.228	2.655	3.057
v/[m^3/kg]	0.000 964	0.000 980	0.001 008	0.001 045	0.001 094	0.001 157	0.001 240
p/[MPa]				**90**			
t/[°C]	0.0	50	100	150	200	250	300
h/[kJ/kg]	86.7	285.6	487.4	691.1	897.7	1108.6	1326.4
s/[kJ/kg K]	−0.005	0.663	1.244	1.756	2.217	2.641	3.038
v/[m^3/kg]	0.000 960	0.000 977	0.001 004	0.001 040	0.001 088	0.001 149	0.001 227
p/[MPa]				**100**			
t/[°C]	0.0	50	100	150	200	250	300
h/[kJ/kg]	95.9	293.9	495.1	698.0	903.5	1113.0	1328.7
s/[kJ/kg K]	−0.007	0.658	1.237	1.748	2.207	2.627	3.021
v/[m^3/kg]	0.000 956	0.000 973	0.001 000	0.001 036	0.001 082	0.001 141	0.001 216

Table A3 *(Cont'd)*

p/[MPa]			22.12			
t/[°C]	300	350	400	450	500	550
h/[kJ/kg]	1332.8	1636.6	2733.4	3019.7	3209.9	3370.2
s/[kJ/kg K]	3.201	3.708	5.401	5.813	6.067	6.268
v/[m^3/kg]	0.000 354	0.001 634	0.008 156	0.011 03	0.013 03	0.014 72

p/[MPa]			**30**			
t/[°C]	300	350	400	450	500	550
h/[kJ/kg]	1328.7	1610.0	2161.8	2825.6	3085.0	3277.4
s/[kJ/kg K]	3.176	3.646	4.490	5.449	5.797	6.039
v/[m^3/kg]	0.001 332	0.001 554	0.002 831	0.006 735	0.008 681	0.010 17

p/[MPa]			**40**			
t/[°C]	300	350	400	450	500	550
h/[kJ/kg]	1325.4	1589.7	1934.1	2515.6	2906.8	3151.6
s/[kJ/kg K]	3.147	3.588	4.119	4.951	5.476	5.784
v/[m^3/kg]	0.001 308	0.001 490	0.001 909	0.003 675	0.005 616	0.006 982

p/[MPa]			**50**			
t/[°C]	300	350	400	450	500	550
h/[kJ/kg]	1323.7	1576.4	1877.7	2293.2	2723.0	3021.1
s/[kJ/kg K]	3.121	3.544	4.008	4.603	5.178	5.552
v/[m^3/kg]	0.001 287	0.001 444	0.001 729	0.002 492	0.003 882	0.005 113

p/[MPa]			**60**			
t/[°C]	300	350	400	450	500	550
h/[kJ/kg]	1323.2	1567.1	1847.3	2187.1	2570.6	2896.2
s/[kJ/kg K]	3.098	3.506	3.938	4.425	4.937	5.346
v/[m^3/kg]	0.001 270	0.001 408	0.001 632	0.002 084	0.002 952	0.003 947

p/[MPa]			**70**			
t/[°C]	300	350	400	450	500	550
h/[kJ/kg]	1323.6	1560.6	1827.8	2129.9	2467.1	2790.7
s/[kJ/kg K]	3.077	3.473	3.885	4.318	4.769	5.175
v/[m^3/kg]	0.001 254	0.001 379	0.001 567	0.001 890	0.002 467	0.003 222

p/[MPa]			**80**			
t/[°C]	300	350	400	450	500	550
h/[kJ/kg]	1324.7	1555.9	1814.2	2094.1	2397.4	2708.0
s/[kJ/kg K]	3.057	3.444	3.842	4.243	4.649	5.038
v/[m^3/kg]	0.001 240	0.001 355	0.001 518	0.001 772	0.002 188	0.002 764

p/[MPa]			**90**			
t/[°C]	300	350	400	450	500	550
h/[kJ/kg]	1326.4	1552.7	1804.6	2069.3	2349.9	2643.0
s/[kJ/kg K]	3.038	3.417	3.806	4.185	4.560	4.928
v/[m^3/kg]	0.001 227	0.001 334	0.001 479	0.001 691	0.002 013	0.002 458

p/[MPa]			**100**			
t/[°C]	300	350	400	450	500	550
h/[kJ/kg]	1328.7	1550.6	1797.6	2051.2	2316.1	2593.8
s/[kJ/kg K]	3.021	3.392	3.774	4.137	4.491	4.839
v/[m^3/kg]	0.001 216	0.001 315	0.001 446	0.001 629	0.001 893	0.002 246

Table A3 *(Cont'd)*

p/[MPa]			22.12		
t/[°C]	550	600	650	700	750
h/[kJ/kg]	3370.2	3516.3	3655.2	3790.3	3923.4
s/[kJ/kg K]	6.268	6.441	6.595	6.738	6.871
v/[m^3/kg]	0.014 72	0.016 23	0.017 63	0.018 96	0.020 24
p/[MPa]			**30**		
t/[°C]	550	600	650	700	750
h/[kJ/kg]	3277.4	3443.0	3595.0	3739.7	3880.3
s/[kJ/kg K]	6.039	6.234	6.403	6.556	6.697
v/[m^3/kg]	0.010 17	0.011 44	0.012 58	0.013 65	0.014 65
p/[MPa]			**40**		
t/[°C]	550	600	650	700	750
h/[kJ/kg]	3151.6	3346.4	3517.0	3674.8	3825.5
s/[kJ/kg K]	5.784	6.013	6.204	6.370	6.521
v/[m^3/kg]	0.006 982	0.008 088	0.009 053	0.009 930	0.010 75
p/[MPa]			**50**		
t/[°C]	550	600	650	700	750
h/[kJ/kg]	3021.1	3248.3	3438.9	3610.2	3770.9
s/[kJ/kg K]	5.552	5.821	6.033	6.214	6.375
v/[m^3/kg]	0.005 113	0.006 111	0.006 960	0.007 720	0.008 420
p/[MPa]			**60**		
t/[°C]	550	600	650	700	750
h/[kJ/kg]	2896.2	3151.6	3362.4	3547.0	3717.4
s/[kJ/kg K]	5.346	5.648	5.883	6.077	6.248
v/[m^3/kg]	0.003 947	0.004 835	0.005 596	0.006 269	0.006 885
p/[MPa]			**70**		
t/[°C]	550	600	650	700	750
h/[kJ/kg]	2790.7	3060.4	3289.0	3486.3	3665.8
s/[kJ/kg K]	5.175	5.493	5.748	5.956	6.136
v/[m^3/kg]	0.003 222	0.003 972	0.004 652	0.005 257	0.005 808
p/[MPa]			**80**		
t/[°C]	550	600	650	700	750
h/[kJ/kg]	2708.0	2980.3	3220.3	3428.7	3616.7
s/[kJ/kg K]	5.038	5.359	5.627	5.847	6.035
v/[m^3/kg]	0.002 764	0.003 379	0.003 974	0.004 519	0.005 017
p/[MPa]			**90**		
t/[°C]	550	600	650	700	750
h/[kJ/kg]	2643.0	2913.5	3158.2	3374.6	3570.1
s/[kJ/kg K]	4.928	5.247	5.520	5.748	5.944
v/[m^3/kg]	0.002 458	0.002 967	0.003 476	0.003 964	0.004 419
p/[MPa]			**100**		
t/[°C]	550	600	650	700	750
h/[kJ/kg]	2593.8	2857.5	3105.3	3324.4	3526.0
s/[kJ/kg K]	4.839	5.151	5.427	5.658	5.860
v/[m^3/kg]	0.002 246	0.002 668	0.003 106	0.003 536	0.003 952

Appendix B

Ideal gas data

The ideal gas parameters given in this appendix are for use in calculations where ideal gas behaviour with constant specific heats is assumed. The specific heat values at constant pressure are those at 300 K and 0.1 MPa. The values of R, c_v and γ are calculated from $\overline{m}$, $\overline{R}$ and c_p.

Universal gas constant, $\overline{R} = 8.3144$ kJ/kmol K

Gas	*Formula*	$\overline{m}$ [kg/kmol]	R [kJ/kg K]	c_p [kJ/kg K]	c_v [kJ/kg K]	γ
Air	—	28.97	0.2870	1.005	0.718	1.400
Argon	Ar	39.948	0.2081	0.5203	0.3122	1.667
Carbon dioxide	CO_2	44.01	0.1889	0.8457	0.6568	1.288
Carbon monoxide	CO	28.01	0.2968	1.0404	0.7436	1.399
Helium	He	4.0026	2.0772	5.1931	3.1159	1.667
Hydrogen	H_2	2.0159	4.1244	14.27	10.1456	1.407
Nitrogen	N_2	28.013	0.2968	1.039	0.7422	1.400
Oxygen	O_2	31.9988	0.2598	0.9183	0.6585	1.395

Principal data source: Chase, M.W. *et al.*, *JANAF thermochemical tables*, third edition, parts I and II, American Institute of Physics Inc., New York, 1986.

Appendix C

Answers to the self-assessment questions

Chapter 1

1.1 33.54 kg, 329.0 N
1.2 6.112 kW

Chapter 3

3.1 147.2 J
3.2 42.88 MJ

Chapter 4

4.1 50 J/kg, 98.1 J/kg
4.2 22.3 °C

Chapter 5

5.1 (a) 34.86 J, (b) 12.24 J
5.2 75.88 kJ, −93.33 kJ, −17.47 kJ
5.3 (b) 1.667, (c) 2.190 kJ, 1.566 kJ, −4.223 kJ, −1.566 kJ, (d) −2.034 kJ

Chapter 7

7.1 (a) 167.5 kJ/kg, 0.110 118 m^3/kg, (b) 0.883, (c) 0.551 m^3/kg,
(d) 254.6 °C, 2600.2 kJ/kg
7.2 (a) 0.961, (b) 0.4111 m^3/kg, (c) 305.0 °C
7.3 (a) 2850.1 kJ/kg, (b) 1213.7 kJ/kg $\leq h \leq$ 2785.0 kJ/kg, (c) 2294.6 kJ/kg,
(d) 421.1 kJ/kg, (e) 1093.6 kJ/kg
7.4 9.655 m^3/kg, 0.8015

Chapter 8

8.1 0.0625 m^3
8.2 0.1339 m^3, 1405.8 kJ
8.3 (a) 0.168 kg, (b) 89.6 °C, (c) 0.089 16 m^3/kg, 0.3567 m^3/kg,
(d) −7.271 kJ
8.4 (a) 722.6 kJ, (b) 182.6 kJ

Chapter 9

9.1 0.014 01 m^3
9.2 −0.0412 kg/s
9.3 0.030 25 kg/s

Chapter 10

10.1 258.5 °C, 204.0 MJ
10.2 728.3 J/kg K, 520.4 J/kg K, 1.40, 208.0 J/kg K
10.3 56.09 kJ
10.4 143.3 °C, 0.399 MPa, 155.1 MJ
10.5 −110.8 kJ, −69.32 kJ
10.6 (b) 1.113 kJ, 0, −2.792 kJ, 0, −1.679 kJ, (c) −1.113 kJ, 2.292 kJ,
2.792 kJ, −2.292 kJ, 5.084 kJ, (d) 81.17 kW, 245.7 kW

Chapter 11

11.1 0.967 MW
11.2 20.35 MW into the steady flow system
11.3 74.37 kJ/kg, 0.444 kW from the compressor to the surroundings
11.4 Yes

Chapter 12

12.1 19.18%, 39.6 kW
12.2 (a) not a heat engine, (b) heat engine, (c) not a heat engine, (d) not a heat engine, (e) not a heat engine, (f) a heat engine operating in reverse: this serves as both a heat pump and a refrigeration plant
12.3 23.43 kW, 64.43 kW
12.4 1.014, 47.1 W
12.5 23.56%, 37.09 kW

Chapter 13

13.1 66.8%, 30.46 kW, 15.14 kW
13.2 2.44 kW, 15.04 kW
13.3 (a) 15.99 kW, (b) 53.3%, (c) the claims do not violate the first law; the claims violate the second law

Chapter 14

14.1 0.415 J/K

14.2 $\Delta s_{1\to2} = c_p \ln\left(\frac{T_2}{T_1}\right) - R \ln\left(\frac{p_2}{p_1}\right)$

14.3 (a) 0.106 kJ/kg K, (b) 6.139 kJ/kg K, (c) 6.218 kJ/kg K

Chapter 15

15.1 69.31 J, −69.31 J, 154.7 J, 0 J, 38.2%
15.2 30.8%, 0.3742 MW, 0.842 MW
15.3 0.806, 0.170
15.4 29.46, 7.61

Chapter 16

16.1 6.64 MW
16.2 32.0%
16.3 4.047 m^3/s, 69.3 m^3/s

Chapter 17

17.1 418.4 °C, 2089.7 °C
17.2 (a) 4.318×10^{-4} kg, (b) 0.415 kJ, (c) 56.5%, (d) −0.234 kJ

Appendix D

Guide to further reading

The most important advice in this section is not about what further references on thermodynamics to consult, but to make a point of looking at some other references. Of necessity, this book provides only a brief treatment of the various topics it introduces. There are very many excellent textbooks available and almost any book on engineering thermodynamics will cover most of the basics that are included in this book. It may be worth while to browse in a library to find a selection of textbooks that you like. You could then refer to them from time to time.

Some contemporary textbooks

Joel, Rayner (1966) *Basic Engineering Thermodynamics in S.I. Units*, third edition, Longman, London. This book is very easy to follow.

Moran, M.J. and Shapiro, H.N. (1988) *Fundamentals of Engineering Thermodynamics*, John Wiley and Sons, New York. This includes a good treatment of the second law.

Rogers, Gordon and Mayhew, Yon (1992) *Engineering Thermodynamics Work and Heat Transfer*, fourth edition, Longman Scientific and Technical, Harlow, England. A comprehensive textbook to honours degree level. Work done on a system is taken as positive.

Van Wylen, Gordon J., Sonntag, Richard E. and Borgnakke, Claus (1994) *Fundamentals of Classical Thermodynamics*, fourth edition, John Wiley and Sons, New York. A well-established modern textbook to honours degree level, including chemical thermodynamics.

Some classical references

Carnot, S. (1824) *Réflexions sur la puissance motrice du feu et sur les machines propres a développer cette puissance* (Reflections on the motive power of heat and on machines fitted to develop this power, translated by S. Thurston, ASME, New York, 1943). Carnot's book is arguably the most important ever written in the field of thermodynamics. It is still an interesting and readable short book.

Clausius, R. (1879) *The Mechanical Theory of Heat*, translated by Walter R. Browne, Macmillan, London. This contains the original Clausius statement of the second law, the original derivation of the Clausius inequality and the original definition of the property entropy.

Keenan, Joseph H. (1970) *Principles of General Thermodynamics*, MIT Press, Cambridge, Massachusetts (originally published 1941, John Wiley, New York). This book is a classic and is still remarkable for its clarity.

Some advanced textbooks

Bejan, Adrian (1988) *Advanced Engineering Thermodynamics*, John Wiley and Sons, New York. This deals with advanced topics of current interest.

Gyftopoulus, E.P. and Beretta, G.P. (1991) *Thermodynamics: Foundations and Applications*, Macmillan, New York. This presents a radical, thought-provoking and scholarly approach to thermodynamics.

Kotas, T.J. (1985) *The Exergy Method of Thermal Plant Analysis*, Butterworths, London. This is an established reference book on exergy analysis.

Index